cowries

Second Edition, Revised

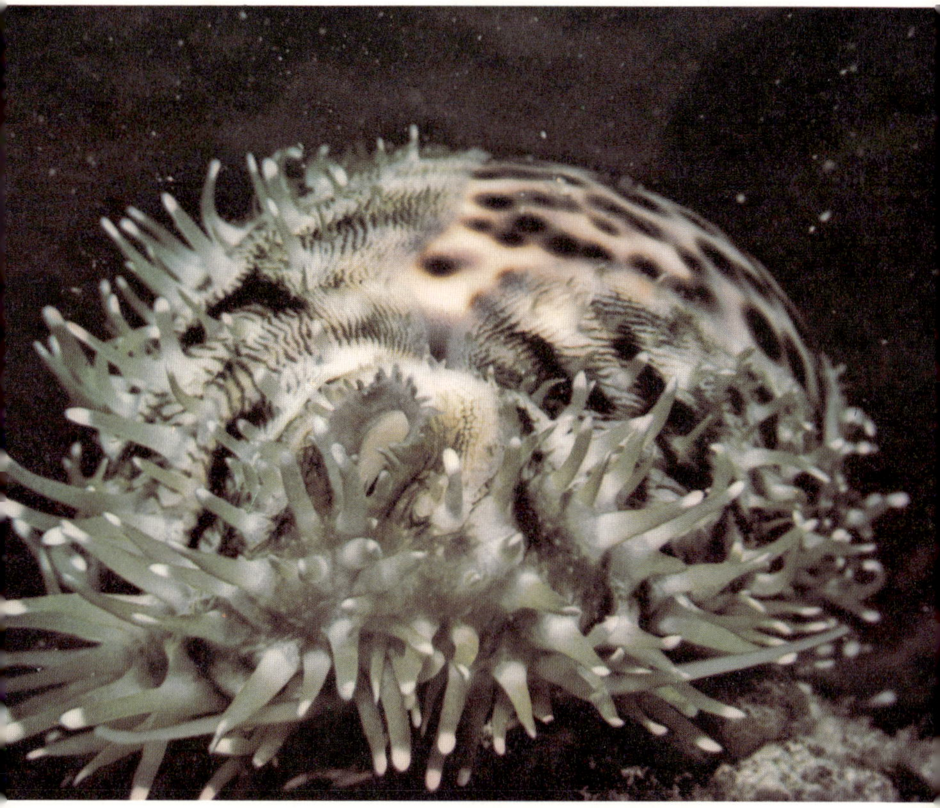

JERRY G. WALLS

With a section on *The Living Cowry*
by Dr. John Taylor

Title page: *Cypraea tigris* with the mantle incompletely exposed, photographed on Heron Island, Queensland, Australia by Jean Deas.

All photos by Dr. Herbert R. Axelrod unless otherwise acknowledged.

The photos of *Cypraea bernardi* are used through the courtesy of Les Editions du Pacifique, photographer Claude Rives, and Dr. G. Richard.

ISBN 0-87666-630-6
© 1975 T.F.H. Publications Inc. Ltd.
© 1979 T.F.H. Publications Inc. Ltd.

Distributed in the U.S. by T.F.H. Publications, Inc., 211 West Sylvania Avenue, P.O. Box 427, Neptune, N.J. 07753; in England by T.F.H. (Gt. Britain) Ltd., 13 Nutley Lane, Reigate, Surrey; in Canada to the book store and library trade by Beaverbooks, 953 Dillingham Road, Pickering, Ontario L1W 1Z7; in Canada to the pet trade by Rolf C. Hagen Ltd., 3225 Sartelon Street, Montreal 382, Quebec; in Southeast Asia by Y.W. Ong, 9 Lorong 36 Geylang, Singapore 14; in Australia and the South Pacific by Pet Imports Pty. Ltd., P.O. Box 149, Brookvale 2100, N.S.W., Australia; in South Africa by Valiant Publishers (Pty.) Ltd., P.O. Box 78236, Sandton City, 2146, South Africa; Published by T.F.H. Publications, Inc., Ltd., The British Crown Colony of Hong Kong.

CONTENTS
THE LIVING COWRY by Dr. John Taylor
    Introduction 7
    The Cowry Animal 7
    The Shell 10
    Food and Feeding 15
    Movement 18
    Reproduction and Development 19
    Predation Upon Cowries 23
    Habitats 26
    Distribution of Cowries 27
    Geological History of Cowries 35
    Cowries and Man 38
    Conservation 42
    Variation and Taxonomy of Cowries 43
COWRIES AND THE COLLECTOR by Jerry G. Walls
    Aquarium Care 50
    Collecting, Cleaning and Curating 51
    Purchasing Shells 59
COLOR ATLAS OF COWRIES 64-257
SYNOPSIS, PRICING GUIDE AND INDEX
    by Jerry G. Walls and Warren E. Burgess
    Synopsis of Cowry Subspecies and Varieties 258
    Checklist and Pricing Guide to the Cowries 274
    Synonymic Index 278

In life, *Ovula ovum* is covered with a sooty black mantle. This species feeds mostly on soft corals. Photo by Allan Power.

An apparently new species of *Volva* from black coral on Lord Howe Island, New South Wales, Australia. This deep water (180 feet) species has a distinctively colored mantle but unmarked shell. Photo by Walt Deas.

PREFACE TO SECOND EDITION

Since **Cowries** was first published in 1975, several new species of cowries have been discovered and many doubtful taxa have been synonymized. This edition attempts to include changes that are felt to be correct wherever possible. A few photos have been replaced by better specimens and valid new species have been added. Many of the identification captions have been rewritten or somewhat modified, and errors known to be present in the first edition have been corrected. A synopsis of subspecies and major varieties of cowries has been included for those collectors who prefer to collect below the species level. The index and pricing guide have also been completely reworked and brought up to date. It is hoped that this new edition will be even more useful to collectors than the first.

Help for this edition was provided by many people, including Richard M. Kurz and Mal de Mer Enterprises, who loaned shells for photographs; Mr. Les Whatmore, who provided photos of *C. lisetae* and pricing information on South African species; photos of the holotypes of *C. maricola* and *C. fernandoi* were generously provided by Jean and Crawford Cate. Mr. Phillip Clover and Mr. Ray Summers provided many helpful comments on the first edition, some of which have been included here. Our thanks are also extended to those who provided specimens and photos also used in the first edition: Mr. Bob Morrison of Morrison Galleries; H. and Y. Azuma for photos of *C. katsuae, C. sakurai,* and *C. teramachii*; Dr. M. Brittan for several photos; Mr. N. Gray for photos of *C. langfordi* and several other shells; and Drs. M. Hyett and H.R. Axelrod for important specimens. Mrs. K. Way assisted Dr. Taylor in preparing *The Living Cowry* chapters.

Finally, Warren E. Burgess helped immensely with his comments for the *synopsis* of subspecies. He also provided the cards used to assemble the new pricing guide and helped check over the corrections on the captions. To him my greatest thanks.

THE LIVING COWRY

Introduction

Cowries are largely tropical, shallow water, omnivorous, epifaunal, largely nocturnal gastropod molluscs. They first evolved about 140 million years ago, and there are nearly 200 species living today. Their domed, highly polished, beautifully patterned shells have tremendous esthetic appeal and for several thousand years have been collected by peoples of widely different cultures. Today with collectors all over the world they are the most popular of all shell groups.

Cowries belong to the superfamily Cypraeacea, which is classified into the larger group of gastropods known as the Mesogastropoda, which among others also includes such well known superfamilies as the strombs and conchs (Strombacea), the tun and frog shells (Tonnacea) and the winkles (Littorinacea). The superfamily Cypraeacea is further divided into three families: the Ovulidae, which includes the egg shells, with genera such as *Ovulum, Simnia* and *Cyphoma*; the Eratoidae, which includes the well-known genus *Trivia*; and thirdly the Cypraeidae, or true cowries, which are dealt with in this book. The shells of members of the Ovulidae and Eratoidae look very similar to cowries, but the animals inside differ considerably and the three families have been distinct for at least 80 million years. The family Ovulidae lives almost exclusively upon coelenterates such as soft corals, sea whips and sea fans, and the shell shape is often highly modified to suit the specialized habitat. Members of the Eratoidae are generally small and live mainly in association with sea squirts.

The Cowry Animal

The main external anatomical features of cowries are shown in the accompanying illustration. The most striking feature is the bilobed mantle which, when extended, covers the whole shell; a sinuous line on the shell often marks the position where the two lobes meet. The character of the outer mantle surface varies considerably from species to species. In some species it is smooth, others have tubercles, some have finger-like papillae and many species have long finely branching papillae sometimes arranged in tufts. The function of these papillae is unknown; they may be

Three common genera of allied cowries to show relative size (right to left): *Ovula*, *Cyphoma*, and *Trivia*.

Ovula ovum, the common egg cowry, one of the allied cowries of the family Ovulidae. The total absence of teeth on the inner lip is only one of many characters distinguishing it from typical cowries.

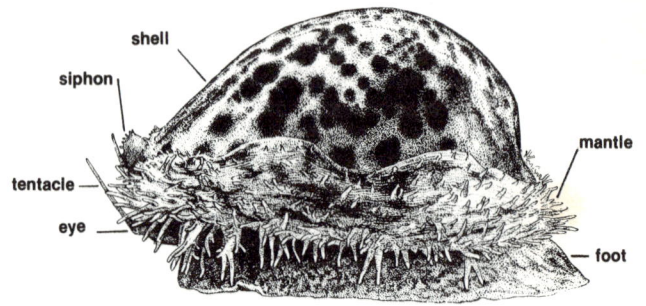

External structure of a living cowry. From *Australian Shells*, by B.R. Wilson and K. Gillett, A.H. & H.A.W. Reed, publishers.

sensory, and some people have suggested they may be concerned with respiration. In some species the mantle is very thin and the shell pattern may be seen through it, but in most cowries the mantle is conspicuously patterned and, although a mottled effect of browns, beige and black is usual, some species show spectacular colors very different from those of the shell. *Cypraea isabella*, for instance, is jet black, *C. talpa* dark gray and *C. teres*, *C. cribraria* and *C. fimbriata* are red. The outer surface of the mantle is glandular and secretes a strongly acid fluid when disturbed.

The siphon, through which water is drawn for respiration, protrudes from the anterior end of the shell; inside the animal at the base of the siphon is the osphradium, which is a chemo-sensory device for sampling the water being drawn over the gill. Below the siphon is the proboscis, which is concerned with feeding; on either side of it are two tentacles which bear well developed eyes. The foot is broad and extensible.

The Shell

SHAPE AND STRUCTURE

The shell of cowries is unusual among marine gastropods because of the domed, almost hemispherical, bilaterally symmetrical shape and the highly glossy outer surface. The shell shape is produced from a normally spiral shell by a combination of the elongate aperture and almost planispiral coiling, so that the shell in growth coils upon itself. The polished surface is produced by the action of the bilobed extensible mantle.

Protection is one of the main functions of the mollusc shell, and some features of the cowry shell are concerned with strength and protection from predators and other environmental insults. The strength of shells results from a combination of the strength of the particular materials the shell is built from and strength factors intrinsic in the geometry of the shell itself. The domed shape of the cowry is an inherently strong structure, and stresses impinging upon the shell will tend to be transmitted through the plane of the shell rather than across the shell, which would tend to cause fracturing at the points of stress. The shape and glossiness of the shell are an advantage in resisting predatory attacks; crabs, for instance, will have difficulty manipulating a cowry in their claws (try simulating this by picking up a cowry with forceps).

As in most other molluscs, cowry shells are built from a combination of two materials: calcium carbonate crystals, in the mineral form aragonite, and an organic component largely consisting of fibrous protein. Many different types of crystal arrangement occur in shells, but in cowries the shell consists throughout of a microstructure known as 'crossed-lamellar.' At microscopic level this is seen to consist of thin needle-like crystals which are arranged into lamellae; in adjacent lamellae the crystals are aligned in opposing directions rather like the grain in each layer of plywood, but repeated hundreds of times over. The cowry shell consists of several layers of this crossed-lamellar structure, with each layer oriented in a different direction. In mechanical terms this means that any crack developed due to loading that causes a fracture in one layer will tend not to be transmitted through the whole thickness of the shell; instead it will be dissipated along the layer boundaries. Of course, in life some stresses, such as in some types of fish predation, are too high to prevent fracture of the whole shell.

The protein component of the shell, which surrounds the tiny crystals, is important during the secretion of the crystals and also contributes toward the strength of the shell. In the same way that fibers of glass mixed with a resin as in fiber-glass is a stronger material than an equivalent sized piece of glass, a finely crystalline mixture of calcium carbonate and fibrous protein will be stronger than an equivalent sized crystal of calcium carbonate.

Major parts of the cowry shell.

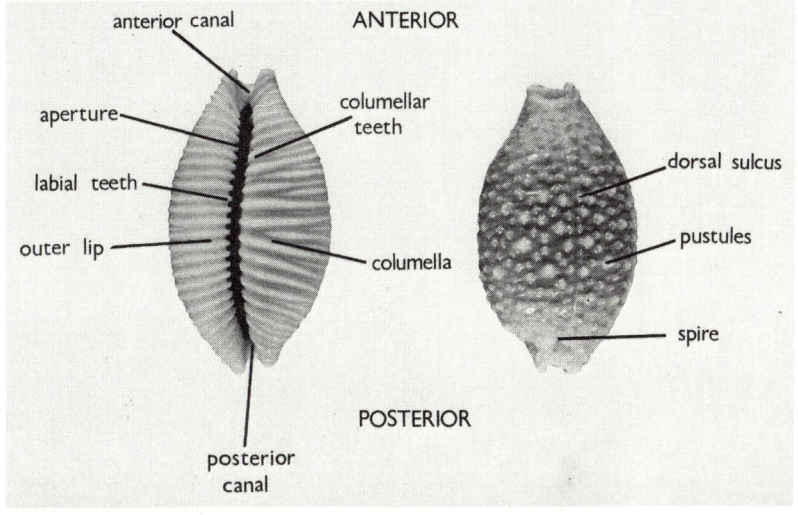

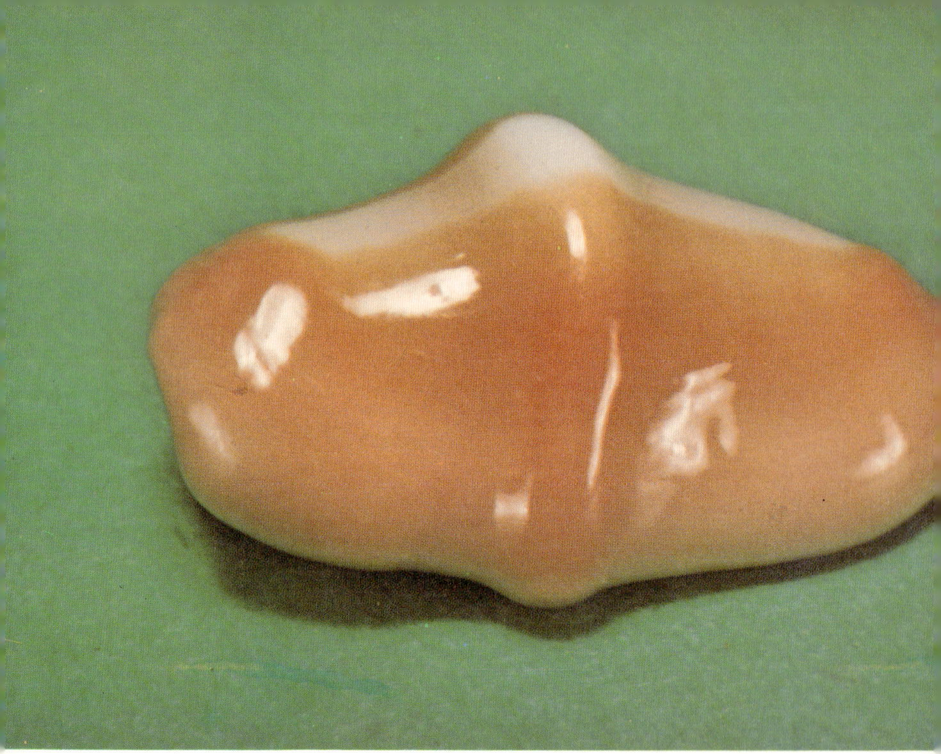

Cyphoma gibbosum is a common Caribbean species of allied cowry. The heavy ridge at the middle of the body and the lack of teeth easily separate *Cyphoma* shells from the true cowries.

Living *Cyphoma gibbosum* are much more attractive than just the shell. Unlike the true cowries, which seldom have variegated color patterns on the mantle, many allied cowries have brilliant spots or stripes. Photo by Aaron Norman.

Apertural teeth are very distinctive features of most cowries, but their function is largely unknown. It could be that, as cowries lack an operculum, an apertural barrier formed by the teeth will prevent or deter the entry of certain predators; the outer lip will also be strengthened by the teeth acting as thickening buttresses. The teeth may also act as a sort of device for cleaning the mantle of extraneous material as it is retracted into the shell.

COLOR PATTERNS

Although one of the most distinctive features of cowries is the beauty and diversity of their color patterns, almost nothing is known of their nature, mode of secretion or significance. The pigments are complex organic molecules often derived from plant food or as byproducts of metabolic processes. The main pigment groups which form the colors in cowry shells are the carotenoids (which usually form the orange and yellow colors), the melanins (which form black and brown colors) and the porphyrins.

The pigments are secreted by cells in the outer surface of the mantle. The pigment molecules are attached to the organic matrix component of the shell which is incorporated between the calcium carbonate crystals. Different arrangements of pigments cells in the mantle and different rates of secretion by the mantle will produce different patterns in the shell. Thus, in some species, for example *Cypraea asellus*, the pigment secreting cells are confined to bands in the mantle. In shells of the *C. cribraria* group the pattern is produced by a lack of deposition at very specific sites. Most often the pigment is secreted intermittently in diffuse blotches as in *C. tigris* and many other species. Some cowries show aberrant pigmentation, and particularly well known are the very dark melanistic forms found in New Caledonia.

Although there are no definitive studies, it is almost inconceivable that the color patterns of cowries do not have adaptive significance; in some species the patterns are obviously cryptic, acting as a camouflage against predators. We must remember that most marine animals do not see colors as we do, and colors seen underwater at night are very different from those seen in the shell cabinet.

The often conspicuously different colors of the mantle and shell may reflect different camouflage needs during the active and quiescent phases of the cowries' daily cycle. It has also been suggested that the surprise of a suddenly withdrawn mantle may frighten off potential predators. In some cases the color of the mantle is very obviously cryptic and species such as *C. teres, C. fimbriata* and *C. cribraria*, which have red and orange mantles, are often found associated with red and orange encrusting sponges. The strongly tuberculate gray mantle of *Cypraea talpa* resembles certain holothurians (sea cucumbers) which are distasteful to many fishes.

The problem that worries many shell collectors is the progressive fading of the color of cowry shells. This is caused by the fact that many pigments are light sensitive and exposure to the ultra-violet wave lengths of light will break down the chemical bonds that hold the pigment molecules together. There is no way of completely preventing the fading, but keeping shells in the dark and dry will considerably retard it.

In many species of cowries the mantle commonly fails to cover the entire dorsum of the shell. This often results in a "mantle mark," a band or stripe of color running from the anterior to posterior end of the shell and interrupting the dorsal color pattern. In some species (for instance *C. mappa*), the band is a constant character, but in most species it may be present or absent, varying from individual to individual. Collectors should not consider the mantle mark as a flaw, but simply as a variable part of the color pattern.

Food and Feeding

As in most gastropods, the cowry feeds by the use of an apparatus unique to the mollusca: the radula. Essentially this is a long strip-like membrane on which are mounted rows of tiny backwardly pointing teeth which are hard and formed largely of chitin. The radula is housed in the anterior part of the mouth, and by a complex system of muscles it can be protruded from the mouth and moved back and forth across a cartilaginous pad

Radula of *Cypraea lynx*. At left is a view of several rows, showing the differences in tooth shape. At right a side view of several marginal teeth from near the end of the rows; these are adapted to gathering food.

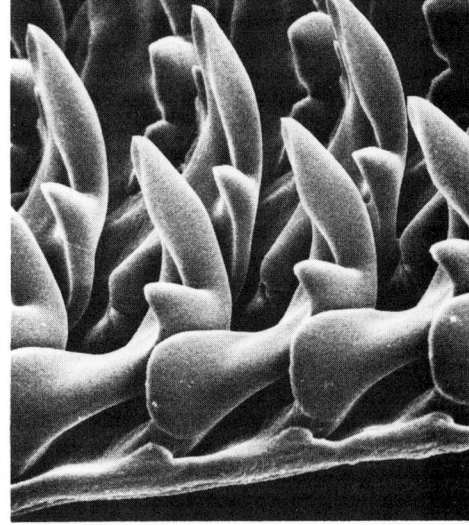

Lateral and ventral view of a *Trivia*, representing the family Eratoidae of the allied cowries. Although the animals are very different, some of the true cowries closely approach this shell shape also.

In cowries the waves of muscular contraction within the foot are easily visible as dark and light lines. This *Cypraea tigris* exhibits diagonal gray stripes on the foot, perhaps indicating that it is about to make a turn. Photo by Roger Steene.

called the odontophore. When the radula is moved forward across the cartilage the teeth become erect and are kept rigid by the locking of the teeth behind them. As the radula moves forward, the tip of the proboscis is applied to a particular food item and pieces of it are gouged off by the teeth. On backward movement the lateral teeth fold inward and the teeth retract, pulling pieces of food into the mouth. This scraping action of the teeth against rock surfaces quickly wears away the teeth, but only about ten per cent of the teeth are used at any one time. As old teeth are worn away at one end of the radula ribbon, new ones are being continually formed at the other.

On each radula ribbon there are up to about 200 rows of teeth and usually seven teeth in each row. The individual teeth are all about the same size but differ in shape, which suggests different functions. The accompanying illustrations show the shape of the teeth in *Cypraea lynx*; the three central rows are used for scraping the food and are equipped with sharp cutting cusps, while the sickle-shaped marginal teeth are used for gathering in the food particles. Throughout the whole genus the characters of the radular teeth are very uniform and vary only slightly in the shape and detailed morphology of the cusps. These small differences in the radular teeth may be useful in distinguishing between similar species. For instance, *Cypraea leviathan* and *C. carneola* can only be reliably separated using radula characters. It must, however, be remembered that radular teeth can be as variable as any other character used in taxonomy.

Surprisingly little is known of the diets of cowries, although many observers undoubtedly have unpublished data from aquarium observations. Some species such as *Cypraea friendii, C. venusta* and *C. teres* are known to feed upon sponges, but most species seem to be generalized omnivores. The following list of the gut contents recorded from *Cypraea cervus* in Florida probably typifies the diet of most species: calcareous sediment, abundant green and red filamentous algae, large quantities of sponge spicules, tubes of polychaete worms, foraminifera, copepods, ostracods, bryozoa and small gastropods. Other species such as *C. annulus* include much more plant material in their diet and should be considered as herbivores. Generally cowries seem to be omnivores feeding upon a wide variety of attached algae and encrusting animals; nevertheless there is a great need of proper sampling and recording of gut contents and food choice experiments to fill in details of this important aspect of cowry biology.

Movement

Cowries move by means of waves of muscular contraction in the sole of the foot, rhythmic waves of contraction passing from one end of the foot to the other or from one side to the other. These waves can be seen on the sides of aquaria as alternately dark and light gray stripes. The parts of the foot not involved in the contraction remain firmly attached to the substratum. In moving there must be a compromise between the strength with which the cowry needs to grip the substrate to avoid being dislodged and the speed with which it needs to move to pursue its bodily needs. The type and variety of foot wave patterns present in cowries is unique among the gastropods and gives them great mobility and flexibility of movement. For instance, by using diagonal waves of contraction the cowrie is able to turn on the spot without lifting much of the foot from the substrate. This reduces the chances of becoming dislodged, which may be important in cowries such as *Cypraea caputserpentis* and *C. mauritiana* which live in habitats of high wave action.

Reproduction and Development

In cowries, as in most other gastropods, the sexes are separate; they belong to the gastropod group which practices copulation, although there have been few observations of this. The penis of the male, which is about a third as long as the shell, is extended and inserted into the genital aperture situated toward the rear end of the female. Sperm can be stored within the female (in a seminal receptacle) and can remain active for at least eleven days.

Egg laying occurs some time after copulation; the female chooses a suitable site and with the mantle fully extended circles around the site. The egg capsules appear through the posterior aperture of the shell and by muscular action are moved along a groove formed by an infolding of the sole of the foot. When the capsules reach the middle of the foot they are attached to the substrate. The number of capsules laid, from 100 to 1500, varies considerably from species to species and also between individuals of the same species. The capsules are embedded in a more or less circular gelatinous mass and are sometimes several layers deep. The individual egg capsules are about 2-4mm long and are variable in shape from squat cylinders to globular box-like or vase shapes; they may be white, cream, pink or purple in color. The capsules have a transparent wall containing several hundred eggs within, set in a clear fluid. The fertilized eggs develop and grow within the capsules. Young cowries, in the form of veliger larvae about 0.2mm in size, emerge from the capsule about 18 days later. During the developmental stages the female cowry guards and tends the egg mass by covering it with the foot, by cleansing it and by aggressive behavior toward intruders.

In most species the veliger larva is about 0.2mm in size with an almost spherical shell coiled about one turn; the animal possesses a foot, tentacles, eyes and an operculum. However, at this stage the most important structure, peculiar to the larva, is the *velum* which consists of two flaps of tissue extending from either side of the larva. The velum possesses rows of cilia, and it is the

Cypraea tigris *is one of the most common large cowries throughout the Indo-Pacifia area and also the most conspicuous. Its mottled mantle with large papillae is very distinctive. This sequence of pictures shows the gradual exposure of the mantle in an unusually lightly colored specimen. Photos by Allan Power.*

As exposure begins, notice that the dorsal-most papillae are contracted and indistinct, while those at the anterior end are erect and expanded.

With further opening of the mantle the color pattern of black blotches on tan becomes more apparent, as does the large number of papillae.

Many cowries do not completely cover the shell, leaving a narrow exposed strip usually to one side of the midline of the shell. Since the shell pigments are largely deposited by glands within the mantle, lack of mantle contact may leave a portion of the shell with a different pattern.

beating of these combined with undulations of the whole velum that enables the larva to swim. This is the dispersal phase of the cowry's life cycle, and the tiny veliger larva can easily be transported by ocean currents. The distance traveled depends upon the length of larval life, which varies from species to species but may last several weeks, and also upon the presence and speed of ocean currents in proximity to the hatching site. In this way many cowry species have been gradually dispersed around the prodigious area of the Indo-Pacific province.

In a few species of cowries on the southern Australian coast the cowries hatch from the egg capsules as young cowries and there is no free-living veliger stage.

The veliger larva feeds upon tiny planktonic algae and grows rapidly; at a certain size and in the correct conditions (which vary from species to species) the veliger sinks to the sea bottom. When it finds a suitable site, the velum is resorbed and the young cowry now crawls around using its foot, feeding with the proboscis and radula.

The shell now changes shape and grows rapidly. At this stage it is very thin shelled, elongate in shape and grows by the addition of new shell material to the outer lip. It does not yet have the infolded toothed lips characteristic of the adult cowry, and juvenile cowries are often confused with other molluscs, such as olives. At sexual maturity the outer lip is infolded, thickened and ornamented with apertural teeth. The cowry now cannot grow any larger, and subsequent shell secretion is solely concerned with shell thickening all over the shell but particularly in the apertural area. Shell secretion takes place both from the inner surface of the mantle extended outside of the shell and from the outer surface of the mantle contained within the shell.

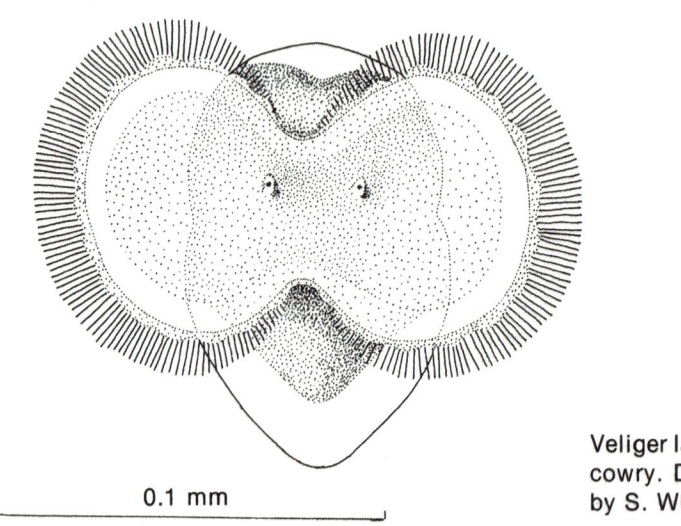

0.1 mm

Veliger larva of a cowry. Drawing by S. Whybrow.

The final size a cowry attains at maturity is genetically controlled but is strongly influenced by food availability and quality and by general environmental conditions during growth. Cowry species vary in size from the smallest *Cypraea minoridens* at around 6-9mm to the large species such as *Cypraea cervus* and *C. tigris* at 125 to over 150mm in length. Considerable variation in size may be found within a species, within a population or over a geographical area. For example, *C. tigris* is significantly larger near the Hawaiian Islands than any other part of its range; similar size differences may be found in many other species.

Within a species there may be size differences between the sexes. In the species where this is known, for example *Cypraea annulus, C. moneta, C. helvola* and *C. angustata*, the male is smaller than the female, although in *C. hesitata* the female is smaller than the male. Environmental conditions affect the adult size of cowries, and populations living in fringe habitats may often be smaller than the main population.

Size is an important ecological character of a species, and differences in size among species which occur together mean that they are potentially able to share the same food but feed upon different parts of it, or share the same habitat but make use of different parts of it, thereby avoiding competition which might result in the exclusion of one of the species. Thus species such as *Cypraea fimbriata*, being very small, can live in small crevices which large species such as *C. tigris* could not possibly enter. There are, however, disadvantages for the small *C. fimbriata* may be exposed to a greater range of predators than large species.

It would be very useful to be able to age cowries; this would enable us to obtain information on the age structure of the population, life expectancy, death rate, recruitment rate, etc. This sort of information is essential if adequate conservation measures are to be implemented for cowries. In many gastropods which grow by addition of new material at the outer lip a record of their previous growth is left incorporated into the shell. By microscopic examination of sectioned shell it is possible to detect diurnal growth increments grouped like tree rings in response to the spring/neap tidal cycle or to summer/winter temperature differences. By this method it is possible to age most marine molluscs; unfortunately cowries do not grow in this way and there is no way yet known for ageing them.

Predation Upon Cowries

It is well known that the behavior of many marine organisms is controlled by fish predation, and it is probable that fish predation is the most likely cause of death in cowries. There are many records of the occurrence of whole cowries in the gut of fishes; the most famous of these records are the South African cowries. The rare *Cypraea broderipii* and *C. fultoni* are known mainly

In *Cypraea cribraria* and its relatives the mantle is usually bright reddish and bears numerous short, blunt papillae. The mantle is thin enough to show the white-spotted pattern underneath, but bright enough to disturb predators. Above is the typical form of the species, with the shell bright reddish brown with large white spots. At right is a specimen of "*Nivigena melwardi*", now known to be a localized albinotic population of *C. cribraria*. Notice the similarity of the mantle in both specimens. Photo above by Roger Steene, that at right by Kieth Gillett.

from specimens found in the stomach of the mussel cracker (*Sparodon durbanensis*). The Brazilian toadfish *Amphichthys cryptocentrus* has a special elongate snout which enables it to probe beneath rock ledges, and several cowry species are regular items in its diet. Many coral reef fish are opportunistic feeders on small invertebrates and will feed upon cowries when available. Quite a number of fish crack the cowry shells before eating, and broken pieces of shell, the result of fish predation, can often be found washed up on the beach.

Crabs are another important predator on molluscs and, although there is very little information to go on, juvenile and adult cowries are probably frequently eaten by the larger predatory crabs.

Other gastropods also eat cowries, and the cone shell *Conus textile* has been reported as eating several species under aquarium conditions, first immobilizing them with poison from the dart-like radula. Some species of *Cymatium* have been recorded as eating cowries.

Habitats

Cowries are dominantly tropical and subtropical animals living mainly in shallow water. Few species live at depths greater than 100 meters; the deepest yet recorded is *Cypraea midwayensis* dredged from 460 meters. While most cowries live on hard substrates of various kinds, be they basalt, granite, limestone, boulders or living or dead coral, there are a few species which often live on soft substrates, usually in shallow water marine grass habitats. These species include *Cypraea mus, C. annulus* and *C. onyx*.

Most cowries are negatively phototropic and hide in crevices and beneath boulders and slabs during the daytime, being active only at night. A few species show the reverse tendency and are active or found in open situations during daylight hours.

The habitats a particular species lives in are largely chosen at the time of larval settlement. In fact, many cowries probably move only relatively short distances during their lifetime. A study of *Cypraea annulus* at Heron Island, Great Barrier Reef of Australia, showed that in a year no individual out of 550 marked had moved more than 5 meters. Factors which may influence the choice of habitat by the settling larva will be the nature and texture of the substrate, the presence of other individuals of the same species and the presence of required food items.

In the tropical Indo-Pacific area as many as 30-50 species may occur at any one geographical locality. In order that so many related species can survive together without competitive exclusions, it is necessary for them to be ecologically different. The usual mechanisms by which species achieve ecological isolation are by habitat differences, food differences or by size; of these, the habitats are the most easily observed. While collecting cowries in an area it soon becomes obvious that some species are very specific and consistent in their habitat distribution, while others are generalists and occur in a wide variety of situations. Species may be arranged with discrete but overlapping ranges along particular environmental gradients such as from shallow to deep or from rough to calm waters. However, most situations are rarely that simple, and the greatest diversity of cowry species is found in coral reef areas where the habitats are geometrically very complex and very variable over small areas.

A section across a typical Indo-Pacific coral reef shows the type of cowry distribution normally met with. The inshore area, although with relatively calm water, is subject to high solar heating, occasional reduced salinities and regular emersion. Toward the edge of the reef the habitat is more complex and often bouldered and with fairly high wave action; the greatest diversity of food items and cowry species occurs here. Further into the surf conditions are more severe and few species are found there. *Cypraea caputserpentis* and *C. mauritiana* are typical of this habitat. Below the surf zone on the reef front where coral is growing profusely, habitats are diverse and a very different kind of cowry species is found.

This example illustrates the manner in which partitioning of habitats occurs. In a particular area careful observation and recording by the collector (or observer) will gradually build up a picture of the habitat distribution of each species and how groups of species interact. For instance, some species may be mutually exclusive while others may regularly occur together but occupy slightly different parts of the same habitat.

Distribution of Cowries

It is well known that cowries are tropical animals, but they do extend from about latitude 40° N to about 40° S, with the greatest diversity of species being found at localities around the Equator. There is a very steep gradient in species diversity between latitudes 20° N and 20° S. The contours of species diversity also show that the Indo-Pacific area is much richer than the tropical Atlantic and Caribbean. This difference is caused by many possible factors, but probably the most important is the much smaller area of the Caribbean.

Individual cowry species are not distributed evenly over the general area occupied by the whole genus. Certain geographical areas have their own distinctive suite of species. Those areas containing the distinctive suites are known as zoogeographic provinces; their boundaries are often vague and transitional, but they form convenient descriptive units. Provinces occur because of the existence of various barriers which prevent the free dispersal of species. They may be physiographical barriers such as land masses or environmental barriers such as sharp discontinuities in water temperature or the direction of ocean currents.

The pale tan mantle of *Cypraea teres* (above) is in striking contrast to the sooty black mantle of *C. clandestina* (below). The mantle of *C. teres* is translucent and allows the shell pattern to show, while that of *C. clandestina* is opaque. The significance of this is unknown. Photos by W. Noel Gray, Guadalcanal, British Solomon Islands.

Cypraea arabica (above) and *C. eglantina* (below), both retracted because they have been frightened by the photographer. Photos by Jean and Walt Deas on Heron Island, Queensland, Australia.

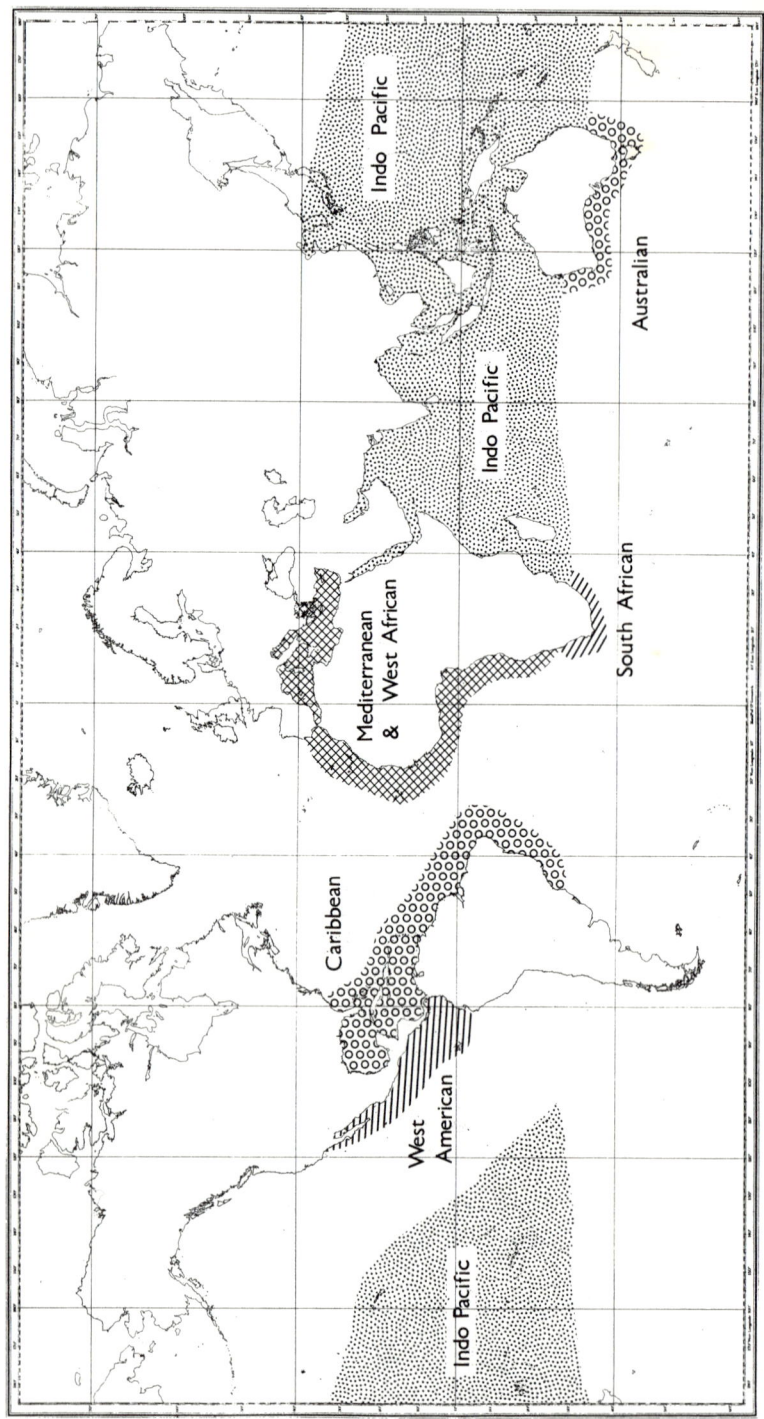

The principle zoogeographic provinces where cowries occur.

Throughout geological time these barriers have been constantly changing, the physiographic ones as a result of the changing position of the continents and the appearance of new island chains, and the environmental ones as a result of global climatic changes. For example, during the height of the last glacial period which ended about 15,000 years ago, large parts of the Indian and Pacific Oceans were too cold for the active growth of reef corals, a factor which must have severely affected the habitats and distribution of many cowry species. Fossil reef deposits dating from the last interglacial period (about 120,000 years ago) in such places as Hawaii contain many cowry species which no longer occur on the islands. Evidence from many places is suggesting that the distributions of species have changed quite considerably, geologically speaking, in a short space of time.

The main zoogeographic provinces containing distinct cowry faunas are illustrated and discussed in more detail below.

THE WEST AMERICAN PROVINCE

This extends from southern California to northern Peru, where it is terminated by the northward flowing cold Humbolt current. Irregular upwelling of cooler water causes the water temperatures in this province to be rather variable. The province supports eight species of cowries; one of these, *Cypraea spadicea*, is found only in the northern sector from Monterey, California to Lower California. Most of the other species, *C. robertsi, C. annettae, C. albuginosa* and *C. cervinetta,* range from the Gulf of California to northern Peru; *C. isabellamexicana* and *C. albuginosa* include the eastern Pacific islands in their range. *Cypraea nigropunctata* is more localized in distribution and is restricted to the coasts of Ecuador, northern Peru and the Galapagos Islands.

CARIBBEAN PROVINCE

This province includes the southern shores of the U.S.A., the Caribbean Sea and the coast of South America to southern Brazil. Four species of cowry, *Cypraea cinerea, C. zebra, C. surinamensis* and *C. spurca*, range throughout the Caribbean from the southern shores of the United States to southern Brazil. *Cypraea cervus* is restricted to the northern Caribbean, while *C. mus* occurs only in a small area on the northern coast of South America. The only cowry to be distributed across both sides of the Atlantic is *Cypraea spurca*.

THE MEDITERRANEAN AND WEST AFRICAN PROVINCES

This area, from the Bay of Biscay to southern Angola and including the Mediterranean Sea, is really two provinces but, because they hold species in common, they are here considered together.

Four species are found in the Mediterranean Sea: *Cypraea spurca, C. pyrum, C. lurida* and *C. achatidea*. All four species extend down the west African coast as far as Angola. The range

Cypraea histrio. Photo by K. Knaack.

Cypraea cervus. Photo by Aaron Norman.

Although mostly retracted, the long black papillae on the mantle of these *Cypraea staphylaea* are still visible. Photo by Jean Deas, north of Heron Island, Queensland, Australia.

of *C. spurca* extends across the Atlantic to the Caribbean; populations on the east side of the Atlantic are more elongate and tend to be brown in color compared to the yellow-orange of their West Atlantic counterparts. Several other species are more truly tropical, and *Cypraea sanguinolenta, C. picta* and *C. gambiensis* are found around the bulge of Africa and the Cape Verde Islands, while *C. stercoraria* and *C. zonaria* extend as far as Angola. Cowries are noticeably absent from the area around the deltas of the great Niger and Zaire rivers, and the relatively low diversity of the tropical Atlantic Coast of Africa is largely attributed to the generally cool currents flowing from both the north and south.

THE SOUTH AFRICAN PROVINCE

The southern tip of Africa is a great mixing zone of oceanic waters, where the warm Agulhas current moving southward from the Indian Ocean meets the cold Benguela current moving northward along the west coast. The Province supports a very distinctive suite of cowries, several of which are extremely rare and known from only a few specimens. Many of the species live in the deeper offshore waters of the continental shelf. The South African species are: *Cypraea amphithales, C. broderipii, C. algoensis, C. capensis, C. citrina, C. cruickshanki, C. edentula, C. fuscorubra, C. fuscodentata, C. fultoni* and *C. gondwanalandensis*.

AUSTRALIAN PROVINCE

The southern shores of Australia are washed by the cool West Wind Drift current, a branch of which flows up the western coast and meets warmer southward flowing currents about half-way.

Eleven species of cowry are known from the province, many of these being very variable species which have caused considerable taxonomic confusion which is now being resolved by adequate biological studies. The group of species similar to *Cypraea friendii* is restricted to West Australia and the southwestern corner of the continent. It includes *C. venusta, C. rosselli, C. marginata, C. friendii* and *C. decipiens*, which occurs only in the northwest part of western Australia. Another group of species are linked in the subgenus *Notocypraea*; five species are usually recognized, with *Cypraea comptoni, C. declivis* and *C. piperita* ranging all along the south coast of Australia and *C. angustata* and *C. pulicaria* being restricted to the southeast and southwest corners respectively. *Cypraea hesitata* is a deep water species of southeast Australia. I consider *C. armeniaca*, found toward the southwest, to be a form of *C. hesitata*.*

THE INDO-PACIFIC PROVINCE

This is much the largest of the provinces. It includes most of

---

* New specimens show that *C. armeniaca* is at least a valid subspecies and possibly a full species. *C. hesitata* is the younger name. —JGW

the Indian Ocean and the tropical Pacific, and its limits extend for a distance of about 23,000km from east Africa in the west to Clipperton and Easter Islands in the east. It includes all of the Indian Ocean north of a line from Durban, South Africa, to about half-way up the coast of Australia and much of the tropical Pacific including southern Japan and the Hawaiian Islands in the north and the myriads of islands in the Pacific. The total area of the Province is about one third of the ocean area of the Earth and totals about 117,400,000 sq. kms, compared with the Caribbean Province which occupies about 5,700,000 sq. kms. Although Japan lies outside of the Tropics, the southern shores are washed by the warm northward flowing Kuroshiro current, which permits the survival of tropical species in such northerly latitudes.

The high diversity of species found in the Indo-Pacific Province is attributed to several causes including: the overall great area of shallow water available for cowries to inhabit; the great variety of habitats available, such as coral reefs, atolls and coral islands, volcanic islands, island arcs and continental margins; the complexity of the habitats such as found in coral communities; and the long term environmental stability, buffered against the severe climatic changes experienced by higher latitudes. Also, the large numbers of isolated islands allow the development of isolated populations and encourages the development of new species.

Such a vast area shows considerable regional variation in its cowry faunas, even though there is a basic homogeneity to the total. As a crude division, 47 species range over both the Indian and Pacific Oceans; a further 25 occur only in the Indian Ocean (including the Red Sea and Persian Gulf) and a further 53 are found only in the Pacific Ocean; another 14 species are found only in the East Indian-Philippine areas. The highest species diversity is found in the western Pacific area.

At the marginal extremities of the Province, very distinctive species with restricted distributions are found. Thus the Hawaiian chain has 7 endemic species, Easter Island 2 and the middle Japanese islands a number of very rare species; at the northern extremities of the Indian Ocean in the Red Sea and Persian Gulf, a further 5 endemic species are found; and southward in Mauritius and Reunion a further distinctive set of species occurs.

Geological History of Cowries

Recognizable cowries first evolved in the shallow tropical waters of the ancient ocean of Tethys in the Upper Jurassic period about 140 million years ago. Later, in the Upper Cretaceous (75 million years ago), they diversified rapidly, and by the early Tertiary (50 million years ago) forms much like those living today had evolved. It is believed that cowries evolved from an extinct group of gastropods called the Columbellinidae, which were probably also the ancestors of the Strombacea (strombs

Cypraea annulus, one of the famous money cowries, is a very common Indo-Pacific species. With its close relative *C. moneta*, it was once a widely used medium of exchange. Photo by Walt Deas.

Opposite page: Above, *Cypraea humphreysi*, more commonly known to collectors as *C. yaloka*. Below, *C. punctata* with the siphon easily visible. Photos by W. Noel Gray, Guadalcanal, British Solomon Islands.

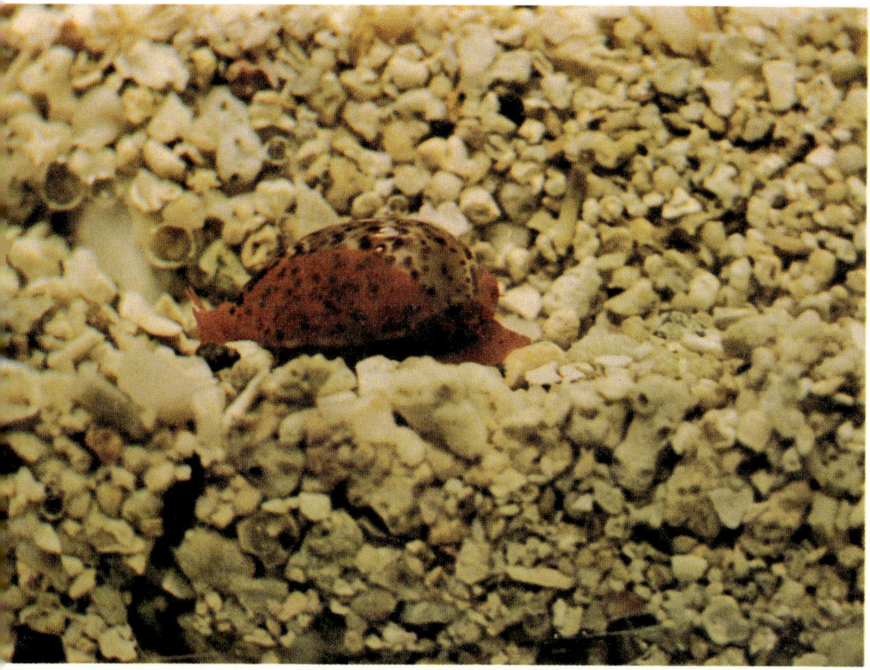

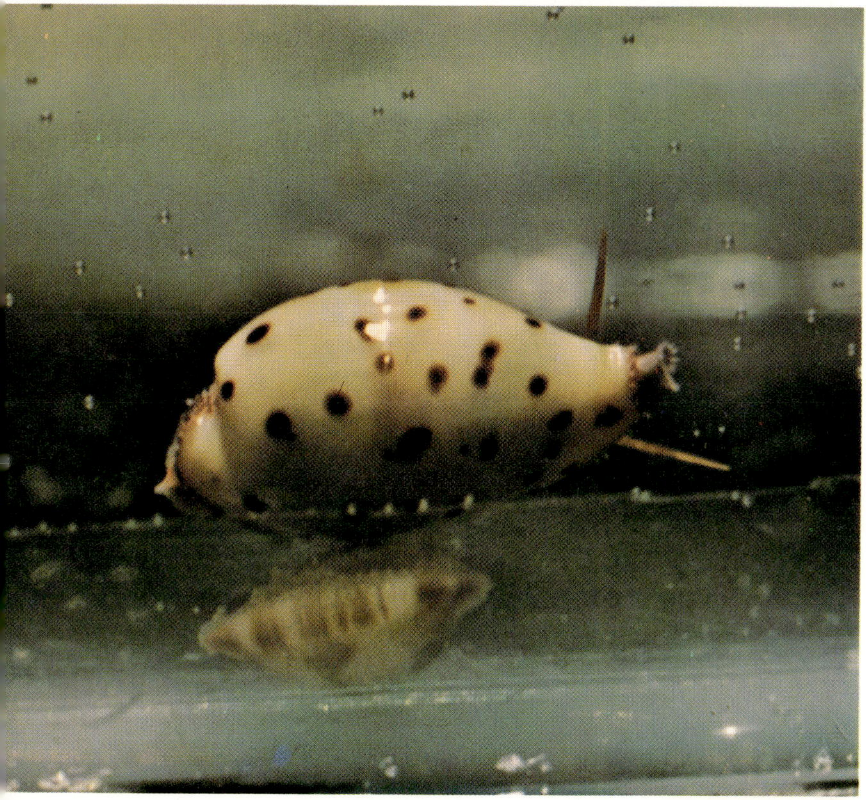

and conchs) and perhaps also the Tonnacea (helmet, triton and frog shells).

The first cowry, called *Zitellia*, had the apertural form of modern cowries with rounded lateral margins and labial teeth. However, the mantle was obviously only partly reflexed and did not cover the whole shell. Later in the Jurassic more normal looking cowries evolved.

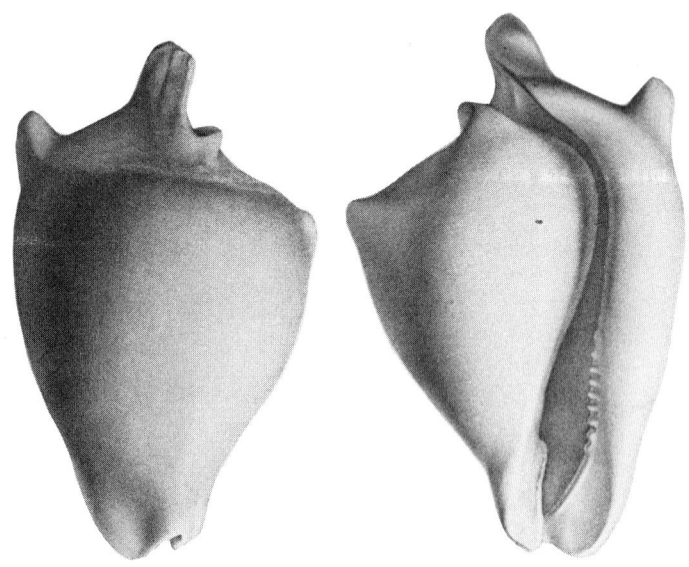

Gisortia gisortiana, an extraordinary fossil cowry from the Eocene (35-55 million years ago) of France.

The most extraordinary cowries were the species of the genera *Gisortia* and *Vicetia*, which flourished in the lower Tertiary period in the Tethys Ocean. They were extremely large, reaching lengths of up to 35cm, and with a thick heavy shell ornamented with large tubercles whose function is a mystery.

Most of the living species of cowry have appeared since the late Miocene (10-20 million years ago).

Cowries and Man

Cowries have been used by man in three basic ways: as ornamental objects; as charms, amulets and talismans; and as a form of money. Their use as ornamental objects is familiar to us today, and cowries are used in many ways as necklaces, brooches, shell "art" works or just by themselves. This type of use has occurred for thousands of years, and cowries are common objects found in archaeological sites of all ages, even those rather distant from where cowries live. The shells are often found pierced, indicating that they were threaded in some way. It is apparent that cowries were collected by earlier peoples with a similar enthusiasm to today's collectors.

More interesting is the belief that was, and still is, current in many cultures, that cowry shells are imbued with talismanic or magical properties. In a variety of cultures around the world cowries have been associated with life and birth, and they were worn on the skirts of girls and women to ensure fertility; in Japan women held cowries during childbirth to ensure an easy delivery. A continuation of the birth association is the practice of putting cowries into graves to confer vitalizing power and ensure an after-life for the deceased.

Cowries were often used as charms against the evil eye, and some authorities have suggested that the resemblance of the apertural side of the cowry to the half closed human eye is the basis for this belief. In India, Persia and Egypt the shells were attached to the trappings of animals such as camels, elephants and horses to ward off the evil eye, and in the Pacific fishermen attach shells to the lines of fishing nets to ward off evil influences. In Egypt, West Africa, Borneo, New Zealand and many other places cowries have been used as eyes in mummies, fetishes and totems and placed in the eye sockets of the dead. In parts of Africa cowries were found as part of the equipment of the "witch doctor" and used in fortune telling, circumcision and baptismal ceremonies.

The most well known use of cowries is, of course, as money. Two species, *Cypraea moneta* and *C. annulus*, were mainly involved; both of them are common and widely distributed in the Indo-Pacific Province. The use of cowry money occurred at some time in many cultures from West Africa to the central Pacific, including the whole of central and eastern Africa, India, Indo-China and China. In some of these places cowry money was a relatively short-lived phenomenon, but in others its use was prevalent up until the beginning of this century.

There are several desirable and necessary properties of materials used as money: they should be handy, lasting, easy to count and difficult to counterfeit. Cowries match these demands well, and they were the most successful and widespread of the non-coinage monies.

It is not known where the practice of using cowries as money began, but they were certainly used in China at about 1000 BC and possibly to a small extent in ancient Egypt. In China their use was superceded fairly early by the introduction of metal coins, but because of frequent forgery of the metal coinage they were revived and continued in use at least as late as the 14th century AD. Interestingly the Chinese word for cowry is incorporated into the written characters meaning wealth, precious and to buy.

Cypraea erosa is a common cowry in many localities. Its grayish mantle bears numerous dendritic or branched papillae which obviously have some camouflage value (photo above) as well as perhaps being distasteful or a nuisance to predators. When the mantle is partially retracted (as on the opposite page) the shell is much more conspicuous against the background. Photos by Jean and Walt Deas.

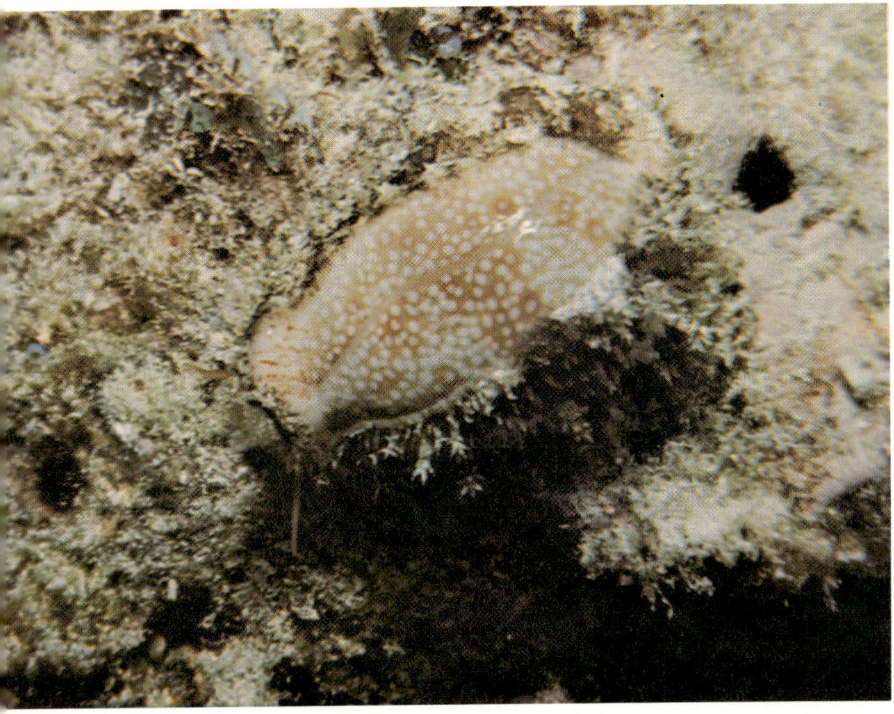

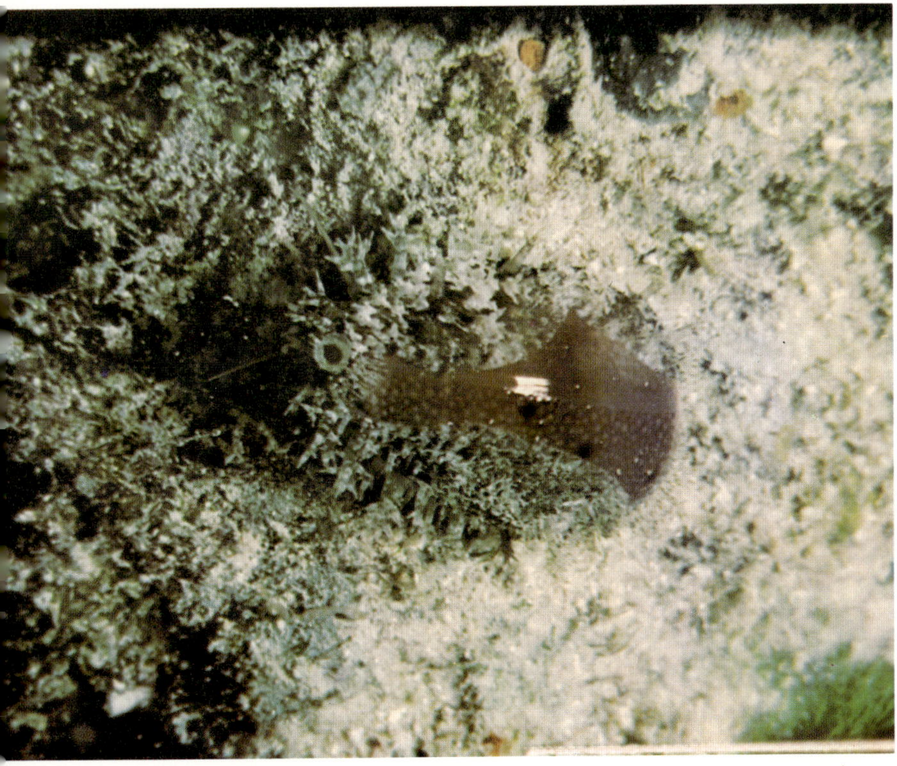

In India the use of cowries had a history of well over 2,000 years and extended well into the last century. As late as 1800 they were the major currency of the district of Silhet, and millions of cowries were in use in this place alone. In Orissa between 1740 and 1845 the exchange for the rupee increased from 2,400 to 6,500 cowries. The main source of cowries was the Maldive and Laccadive Islands, where they were collected by using masses of coconut leaves and branches lashed together and tethered to the sea bottom in shallow water; cowries were attracted to the artificial habitat and crawled between the branches. After a time the branches were dragged ashore and the cowries removed.

In Africa, cowries were in common use during the latter half of the nineteenth century, and the early European explorers of eastern and central Africa such as Burton, Speke, Livingstone and Stanley carried cowries among their array of trading items. However, cowry money was known from much earlier times, and certain Arab records record their use on the east African coast from at least the first century AD. Travellers on the west coast of Africa record cowries being in use during the fourteenth and fifteenth centuries; the cowries were probably introduced by the Arabs for trade and were carried from the east *via* north Africa. In the seventeenth and eighteenth centuries they were often used by slave traders on the Guinea and Senegal coasts. Cowries in east Africa became greatly devalued by over-collection on the coast, a sort of inflation by producing more and more money with no controls. Their value increased in proportion to the distance from the source area.

Conservation

In many areas shell collectors are finding that cowries are becoming increasingly scarce, and it is certain that the rising popularity of shell collecting has contributed greatly to this decline in numbers. With the increasing ease of air travel over the last twenty years, many areas previously rather inaccessible have been "opened up" by tourism, and now large numbers of people spend their vacations in virtually all parts of the tropics. The shallow water inshore habitats in which cowries flourish are highly vulnerable to disturbance and are now under great environmental pressures. Shell collectors are only partly responsible for these pressures, for they come mostly from habitat destruction, discharge of effluents from sewage and industrial processes and silting from soil run-off resulting from intensive agriculture.

On the credit side, shell collectors have often been in the forefront of conservation proposals for the protection of marine faunas and habitat, and, as a result of pressure, the necessity for the designation of protected areas and reserves is becoming increasingly recognized. Indeed, in several countries marine parks and reserves, where any disturbance of wild life is prohibited, have been in operation for a number of years.

One problem of marine parks is that although they are excellent for viewing corals and fish in undisturbed conditions, it is virtually impossible to observe many marine invertebrates such as cowries because of their cryptic habits. The parks serve as refuges that hopefully will harbor and support a reservoir population of cowries that will reproduce and replenish areas outside the reserves.

Compromises are essential since it is obviously impossible to preserve in an unspoilt state all inshore areas of the tropics; equally, collectors and dealers must exercise restraint and control their acquisitive instincts to a reasonable level and at the same time respect and care for the habitats in which they collect and make all efforts to avoid the wanton destruction so prevalent in the last few years.

Shell collectors are usually very observant and often detect subtle environmental changes not obvious to the casual observer; experience has shown that it is by early and firm action that successful measures can be mounted to deal with the source of a disturbance. It is the duty of the shell collector to preserve as much information as possible with his specimens; this should include the exact collection locality, the date of collection, name of collector, information on habitats such as depth, substrate type, position, abundance and so on. This recording of information provides an invaluable record of the history of a fauna in an area and long term changes in species composition, abundance and extinctions. New arrivals can be dated from carefully documented collections. Specimens for which the collection date is known have recently been used to study long term changes in the concentration of trace elements such as mercury, cadmium and zinc.

The great problem in the recommending of conservation measures, for cowries in particular, is the great dearth of essential life history and biological information on which the measures should be based. The information necessary includes: the age structure of populations; the recruitment rate of new individuals to the population; and death rates. Careful observations by collectors could contribute greatly to this necessary background information; careful research and observations of the living animal in the field or aquarium are much more satisfying than amassing drawers and drawers of dead shells, at least in my opinion.

Variation and the Taxonomy of Cowries

Some species of cowries show great variation in morphological characters; these are the species which cause the greatest taxonomic difficulties. However, all species vary in overall size and shape, color pattern of shell and mantle, number and shape of the apertural teeth and shape of the radula; in fact, they vary in all characters. Variation is seen within a local population where no two individuals will be exactly alike, but the population

With the mantle fully withdrawn, this *C. erosa* looks like the proverbial "sore thumb" against its background. Photo by Jean Deas.

On a pale background such as this one, it is probably to the animal's advantage not to expose the mantle. *Cypraea clandestina*, photo by Walt Deas.

Cypraea chinensis has a red mantle with unbranched but very rough papillae. Photo by W. Noel Gray.

may be described statistically by an average set of characters. Species are, however, composed of many local populations. These are adapted to the particular environment in which they live, but environments vary over the total range of the species, and thus the species will show geographical variations. Often characters such as size vary gradually over a species range. This is well known for a number of forms. For example, *Cypraea tigris* becomes progressively larger eastward from the center of its range, and the population of large individuals found in Hawaii is at the extreme end of this size gradient. Morphological response to environmental variables also occurs with increasing depth as, for instance, in the populations of *Cypraea friendii* from deeper water which are both wider and paler than those in shallow water.

Toward the margins of a species' range conditions are, for that species, more extreme, and the populations adapted to these local conditions will here tend to show the greatest divergence from the average characters of the species. Sometimes a population or group of populations of a particular species become isolated and are prevented by some barrier from free genetic exchange with other populations of the same species. If the population remains isolated for an extended period of time, it becomes increasingly adapted to the local conditions and may eventually become sufficiently genetically and ecologically distinct to reinvade the range of the parent species and remain distinct from it without interbreeding. This is the process by which it is believed most new species arise. New species arise from geographically isolated populations, but not all geographical isolates produce new species. Most of these isolated populations are ephemeral, and usually they re-establish contact with and merge back into the main body of the species or they become extinct. Marginal populations are thus of great evolutionary interest.

As far as is possible the concept of a species must take into account the variation in all the populations over the entire range; it is obviously difficult in practice to distinguish between a marginal population of a species, a geographically isolated population, an incipient species and a newly formed species.

These problems are illustrated in the group of species near *Cypraea cribraria*. This species is widely distributed in the Indo-Pacific Province, but the obviously closely related forms *Cypraea cribellum, C. esontropia, C. gaskoini, C. cumingii* and *C. haddnightae* have narrow restricted distributions around the extreme edges of the range of *C. cribraria*. Are these "species" marginal populations of *C. cribraria*, subspecies or full species? The question is probably not resolvable with present knowledge, and more careful and quantitative studies of morphological variation coupled with anatomical and ecological studies are needed. The new powerful tools of biochemical taxonomy may help enormously in unravelling this kind of problem.

Because of their great fascination to collectors, the nomenclature of cowries has suffered more than most molluscan groups from the excessive introduction of specific, subspecific and varietal names. For less than 200 valid species of cowries, well over 1,000 names have been proposed. New species were, and still are, erected on the basis of minor shell morphology and color pattern variations with no regard to or assessment of local or geographical variation. However, an increasingly biological concept of the species has developed among cowry workers so that many names have now been synonymized into more broadly conceived species. As more knowledge becomes available it seems likely that in the future the number of cowry species will be further reduced from the present number; a number of the so-called species recognized today will turn out to be peripheral populations, geographical isolates or subspecies. However, a biologically adequate revision will not be possible for many years.

Some cowry species are known from only one or two specimens and it is obviously impossible in these cases to have any concept of intra-specific variation and how this relates to other closely related species. It is thus difficult to make any real assessment of the status of these species.

Subspecies names are widely used for cowries, and these cause the greatest confusion; too often these names have been given to local variants with no consideration given to the necessary geographical isolation from other populations of the species. In most cases our knowledge of geographical variation within the species is too inadequate to use the subspecies concept in a biologically satisfactory manner. In view of these difficulties we have avoided the use of subspecies names in this book. In some cases forms now accepted as full species are almost certainly only subspecies of more widespread species.

GENERIC NAMES

Many attempts have been made to arrange the species of cowries into groups of similar or related species, and the generic and subgeneric names given to the groups are frequently used in molluscan publications. Some workers have questioned the validity of these generic groupings, pointing out that many morphological characters cross-cut the so-called generic divisions and, indeed, the groupings vary according to what characters are used to sort them. That groups of similar species exist is undisputed, but until we have adequate biological knowledge, for the present time less complication and confusion is caused by merely using the name *Cypraea*.

This well posed *Cypraea vitellus* shows the foot, the dendritic papillae, eyes and eye stalks, and the siphon. Photo by Allan Power.

COWRIES AND THE COLLECTOR

Aquarium Care

Many species of common cowries are very adaptable to aquarium conditions and will survive for years in a well-managed system. The best choice of species is the medium-sized or larger omnivores such as *Cypraea cervus, C. tigris, C. spadicea, C. mappa,* etc. Actually, the smaller species (*C. ziczac, C. minoridens,* etc.) do well in the aquarium, but they are more subject to possible predation from hungry fishes, echinoderms, crabs, etc. For best results, keep cowries and other invertebrates in one tank, fish in another.

Since most cowries are tropical or subtropical in distribution, there need be no great concern over water temperature and salinity—if your marine fishes are surviving well, the conditions will probably also suit the cowries. Very acid water may result in shell erosion and death. It is necessary, however, to provide sufficient hiding places for the use of the snails during the day. Most cowries are active only at night or in subdued light, spending the day hidden under rocks or in small rock and coral caves. Larger fishes which actively prey on snails should not be placed in the same aquarium, nor should large crabs. Although the slippery shell surface, high shape and probably distasteful mantle reduce the chances of a cowry falling prey to a fish, accidents do happen.

Feeding most cowries is not a difficult chore, as many species are omnivores. These will do well on small pieces of finely diced shrimp or fish placed near their daytime hiding place. Be sure that small fish or shrimp do not get to the food first! As a general rule cowries are not great wanderers in the aquarium and they would probably not be able to find food placed more than a few inches away. Since they are nocturnal, it is best to feed at night or when the lights are off. The smaller cowries are often herbivores, satisfied with munching on the green algae growing on the rocks, glass, etc., and supplementing their diet with occasional small crustaceans and other animals which they happen to find in the tank.

Some of the Australian cowries (*Cypraea venusta, C. marginata, friendii,* etc.) are more highly specialized in diet, feeding mainly on sponges. These are of course difficult or impossible to maintain in the average marine aquarium, but it is also very unlikely that such rare and deep water cowries will be available in living condition for the aquarist. There is also a possibility that other cowries have similarly specialized diets, but little is known in this regard. If you do not notice your cowry actively feeding within a couple of days after being placed in the tank and offered both minced seafood and algae, it is best to presume that it has a specialized diet which you will probably not be able either to discover or duplicate.

Occasionally a female cowry will lay several egg capsules in the aquarium, attaching them to a smooth rock or other hard substrate. In the average aquarium it is a lost cause to try to raise the eggs because of the complicated life cycle of the young cowry. It will hatch as the almost microscopic veliger stage, which is normally planktonic in the wild and probably feeds on microscopic algae. After a few days of the proper diet it will transform into a more snaily-looking creature and settle to the bottom. At this point it usually dies in the aquarium, as there appears to be something missing—either the proper food (which is unknown) or the chemicals necessary for successful shell growth.

Never collect a female cowry "setting" on its egg mass or collect the egg cases themselves. In nature the female guards the eggs against smaller predators and helps ensure the survival of as many eggs as possible. Removing the parent even temporarily usually leads to instant death for the eggs, usually by a passing crab or fish.

If properly cared for in a well-maintained aquarium, especially adaptable individual cowries will survive and even grow for several years. *Cypraea cervus* has lived in Florida aquariums for over three years, and there is no reason why many other species should not be as successful. After becoming accustomed to tank conditions, many cowries will become more active during the day, remaining in the open and displaying to all their often fantastically beautiful mantle.

Collecting, Cleaning and Curating

COLLECTING

As explained earlier, cowries are reef animals, and most species are to be found on old reefs with an abundance of dead corals and loose slabs of coral rock. The vast majority of cowries require no specialized collecting methods and can be collected by simply turning over the slabs and pieces of debris during the day. Exercise caution in this, however, as dangerous eels and invertebrates may also be under the rocks. The slabs not only provide shelter for the cowries, but also a substrate on which to lay eggs, a breeding ground for many small crustaceans and other animals used for food, and a surface on which algae grow. Thus failing to return a rock to its original position is a deadly sin—you not only deprive the cowry of its food and shelter, but also numerous other small inhabitants are exposed to the mercy of the reef. *Always leave the rocks and debris in the same position as you found them. Never* destroy any part of the reef by removing the debris,

Above, *Cypraea boivinii*; below, *C. gracilis*. The differences in color and texture of the mantle are striking. Photos by Y. Takemura and K. Suzuki.

Large tiger cowries are often cut and used as napkin rings and other unusual novelties.

breaking corals, etc., as it may be many years before such important habitats are replaced—or they may never be replaced and the reef may die.

Although most cowries can be found in shallow, protected waters easily accessible to the wader, several rarer species and the largest individuals of some species are to be found only by diving. SCUBA divers have the advantage over free-divers because they are able to stay under longer and have more time to check every little crevice and cave. Few cowries are restricted to waters over 50 or 60 feet deep, anyway, and in many cases the dangers of deep diving may begin to outweigh the advantages of collecting for all but the most accomplished divers. Add to this the fact that most cowries can be seen easily only at night when they are feeding in the open, and the necessity for extreme caution and great expertise becomes even more apparent.

A few cowries are apparently restricted to water beyond the normal limits of SCUBA and are found only in the debris of shrimp trawlers and other commercial fishermen. The hard, often jagged bottoms most liked by cowries are dangerous to expensive nets and are avoided by commercial fishermen whenever possible. Thus the apparent rarity of many deep water species in collections. Most of the larger and rarer cowries from Australia and the China Sea are the by-products of commercial trawling and dredging operations.

An interesting sidelight to general cowry collecting is the examination of fish stomachs for fresh and intact cowries. There are many stories of rare cowries and other shells being found in stomachs, and many of these are true. Of course one should look mostly or exclusively in the stomachs of fish which habitually "gulp" molluscs in one piece and don't crush them first. Good hunting might be found in the groupers, snappers, some bottom-dwelling sharks and even in starfishes. You might not find many cowries, but there is always the possibility of finding other types of shells which are almost as interesting.

Whenever living shells are collected, the collector should limit himself to a small series of just the species he needs. Juveniles and brooding females should not be collected. Overcollecting spoils the fun for everyone else—including yourself if you collect in the same area later.

CLEANING

Contrary to the directions in many general shell books, *never* boil a cowry or place it in boiling water. Sudden heating will cause the layers of the shell to expand at different rates, resulting in horrendous cracks. Hot water can also discolor and dull the polished finish of the shell.

If you collect the cowry alive or in freshly dead condition, it is usually very simple to pull the animal out with a knitting needle, hook or bent piece of wire. Difficult specimens can be placed on the sand in a safe place protected from flies and other pests, including humans, and allowed to rot. Be very careful that the shell is turned aperture up so none of the decomposing flesh can get on the outside of the shell. Decomposing flesh is acidic and will rapidly etch into the surface of the shell, causing dull or colorless areas. When enough of the animal has decomposed, the shell is rinsed out with a high pressure jet of water and stubborn pieces pulled out with a wire. *Never* leave the shell uncleaned in water overnight or for more than a couple of hours, as this will also result in acid etching from decomposition products.

The freezer method is a very safe cleaning technique when the facilities are available. The cowry is placed in the freezer for a couple of days, carefully protected against chipping with a plastic bag. It is then thawed thoroughly for a day, being careful that the body does not touch the outside of the shell. The cycle is repeated twice, loosening the muscles which hold the body in the shell. Usually a jet of water and a little picking with a wire will simply and speedily remove the animal after the second thawing.

As a last resort cowries can be preserved in 70% ethyl alcohol, which allows the radula and soft parts to also be studied later. This method should be used cautiously, however, as some brands of alcohol contain denaturants which can impart a dull finish to the shell. Of course formalin solutions, which are acidic, should never be used to preserve any type of shell.

Cowry collectors are lucky because the cowry mantle usually prevents the accumulation of "calcium crud" so common on other types of shells; there are also seldom problems with barnacles, algae, worm tubes, etc. that plague collectors of other shells. After the cowry animal is removed, a thorough rinsing in running water and a careful brushing with a stiff toothbrush should be sufficient to remove minor surface dirt. Calcareous matter could be removed by careful soaking in a weak lye solution, but it is much better to never collect cowries with such defects. Even if removed carefully, the worm tubes and crud will leave bad flaws on the otherwise polished surface.

CATALOGING

Anyone planning an extensive collection of cowries or other shells—extensive meaning anything more than a couple of beach shells to stick on the coffee table—must give careful consideration to an efficient housing and cataloging system for his collection.

The first essential is to see that all data pertinent to a collection (one or more shells of a single species taken at the same time and place) are recorded and kept with the collection. As a rule the basic data *must* include at least the locality where collected (being as specific as possible), the depth and type of bottom, the date of collection, and the collector. Identification of the shell is not really a part of the data, as it can always be added later and is anyway subject to change with the opinions of different people.

A small waste basket made entirely of money cowries. Its great weight assures a stable base.

A handbag made of money cowries and trimmed with *Cypraea caputserpentis*. Such novelty uses of cowries are generally limited in scope and do not draw heavily on the abundance of the species used. Most of the shells probably come from beach specimens which died of natural causes.

Any other data you wish to record, such as tide, associated species, mantle color, etc., are of course welcome.

I find the most simple system consists of a label to be kept with the specimen and an index card for each species. The specimen label should be small enough to put in the box with the shell and yet not small enough to be lost. Cowries are nice in that a folded label can be placed within the aperture where it is available and cannot be lost or mixed up with the label for another shell.

The index card serves as a cross index and can be as complicated or simple as you wish. Use one card for each species (or variety) and simply list on it each collection of that species as it is added to your collection. Thus by looking at the card for *Cypraea annulus*, say, you can instantly see that you have a total of eight collections, three from the Philippines, one from Australia, and four from the Solomons. It is to your advantage to number each collection and record the number on the index card as well.

I doubt if many collectors ink collection numbers and locality on the actual shell anymore. To most people it is disturbing to acquire a shell with someone else's number recorded in a conspicuous place. Besides, polished shells like cowries do not take ink very well, and most people would probably think of an inked shell as being defective. Leave the inking of numbers to large museum collections.

HOUSING

Most books suggest that shells be stored in special cabinets with large, shallow trays, each tray containing numerous smaller cardboard trays, one for each collection of a species. This may sound fine in theory, but in today's world of small apartments and expensive wooden products, a cabinet is simply beyond the reach—or room—of most moderate collectors. In the same way, cigar boxes and other such haphazard containers should not be used unless you are planning to keep only a very few shells.

I find that two housing alternatives are most practical on limited budgets. One is to buy a used chest-of-drawers with the shallowest drawers available, preferably not over three inches deep. This will serve as your basic cabinet and will house a fair number of shells. Individual collection trays can be cut from heavy-weight posterboard and the edges taped with white plastic tape. If the boxes are made in uniform sizes the drawers will look neater and some space will be saved. Boxes an inch deep are best; I use boxes of the following dimensions with good success: 1½x1x1", 3x1x1", 3x2x1", and (rarely) 6x3x1". The first three sizes are the ones you will use most often. Plastic vials with cotton can be used for the very small species. A good grade of cotton fibers or soft tissue can be used to line the bottom of the trays; be sure that if you use colored paper or cotton the dyes are waterfast, or you will get stains on the shell.

An alternative method is to purchase several of the plastic cabinets used for the storage of small parts. These come in many different sizes and designs, from simple boxes partitioned for use with fishing tackle to metal-framed cabinets holding fifty or more plastic trays about 6x3x1" in size; these trays can usually be partitioned into up to four smaller sections. Such a cabinet can hold a large number of small to moderately large specimens with relative safety and at low cost. Large specimens will of course still require some type of larger storage cabinet, but there are fortunately very few cowries too large to fit in one of the standard plastic trays.

GENERAL CARE

Shells are rather permanent objects and will survive many hardships which would destroy lesser objects. But to maintain their interest to collectors, a few simple precautions are essential.

First, keep all shells away from direct sunlight. Bright light of any type, but especially the sun, will cause the rapid fading of bright colors, especially the golds and oranges. Keep the cabinet drawers closed when not in use and cover any transparent plastic tray fronts with paper. If you put a shell on display, make sure it is not placed near a window.

Eventually colors will fade and dull regardless of how the shell is kept, but this is to be expected. Dulled surfaces can be somewhat restored with a *very light* coating of mineral oil; an excess will dry into a gummy residue and make unsightly patches. Be sure that all shells are protected from dust, although a little dust will not hurt and can be removed with a soft tissue.

If several specimens are placed in a tray, be sure they are not able to bump against each other, resulting in chipping, which reduces the value of the shell to collectors. Be especially careful with trays containing large and small specimens of the same species, especially fragile juvenile shells.

Second, shells should never be subjected to extremes of temperature or humidity or sudden changes of temperature. Try to store your collection in a room with low humidity and an even temperature in the seventies. High humidity can result in the growth of fungus on the shells (this can be removed by brushing under running water or by careful application of a weak chlorine bleach with an artist's brush). Very dry climates, however, can hasten fading of the shell.

Purchasing Shells

It is unfortunate that relatively few shell collectors are able to personally collect all their own shells. Even those who live on tropical islands will soon find that there are many species needed for their collection which do not occur in the area or which they

Juvenile cowries (left) differ in several ways from adult shells (right). This photo shows the relatively conspicuous spire of the juvenile, which is nacred-over in adults of most species. The color pattern is seldom as well defined in juveniles as in adults, and important spots, bands, etc. used in identification may be lacking. The teeth seldom are well developed until maturity is reached, so young cowries may have an entirely different appearance in ventral view than would an adult of the same species.

Cypraea hungerfordi, variety or subspecies *coucomi*. This is the Australian representative of a typically Japanese species. Most shells are beach like this one and do not show the typical *C. hungerfordi* pattern very well. Perfect specimens are almost identical with Japanese shells.

are never able to collect themselves. Many will be able to join a local shell club and can trade shells with friends and acquaintances in other countries. But there are still many collectors who will probably never see a living cowry—these collectors must depend on professional shell dealers to provide their wants at a cost.

Many differing opinions have been expressed about shell dealers in general, both pro and con. If you wish to have a complete or nearly complete collection of cowries, or any other shell group for that matter, you will almost certainly have to purchase some of your shells from dealers.

It is only sensible to practice comparative pricing when shopping for shells either at shell shops or through the mail. Most dealers issue free price lists detailing their stock, the lists being published anywhere from bimonthly to yearly. There are many mail-order dealers in the United States and also quite a few in the more exotic areas such as the Philippines, Israel, Reunion, French Polynesia, Australia, etc. Your best bet is to get a handful of price lists from various dealers, read them carefully, and then place a small ($10.00 maximum) order with the dealer(s) you choose. There are bad dealers who will rob you and cheat you, but they are very few and far between. After dealing with a few professionals, you will soon find that one or two can satisfy almost all your demands and provide good specimens at reasonable prices. Find a good dealer and stick with him until a better one comes along—if one does. Good dealers not only provide shells, they can also provide a lot of advice for the more advanced collector on choice of shells, current market values, etc.

Personally, I would never order from a dealer who does not describe the condition of the shells (beach worn shells *must* be described as such) and state their size. The most exact grading terms used are: GEM quality—a perfect specimen, with an unblemished spire, unbroken spines and lips without chips, fully adult and normally colored. A shell without a flaw. Well cleaned, both inside and out. Cowries must have original color and gloss. Bivalves must have both valves, properly matched and unbroken. FINE quality—an adult shell with only minor flaws and with not more than one shallow growth mark. Must have original color and gloss. A cone may have a rough lip OR one small chip. The spire must be unblemished. A *Murex* may have not more than two minor frond breaks. No repairs—filed lips or mended knobs, for example—permitted. GOOD quality—a reasonably acceptable shell, with a few defects such as growth marks, broken spines, worn spire, or lip chips. Minor fading of color permitted. Specimen may be slightly subadult. A good quality shell must faithfully display all the basic characteristics of the species. FAIR quality—may be obviously dead or beach collected with chipped lips, faded color, growth faults or imperfect spires. This grade—comparable to the present "commercial" quality—is not acceptable for mail order retailing and should not be offered as collectors' specimens. (Hawaiian Malacological Society.)

It must be kept in mind that some types of shells, including some cowries, live in areas subject to high wave activity, heavy eroding action of the substrate, or metallic stains in the water; such species never appear in true gem condition, and the fine condition for one species may be equivalent to the gem condition for another.

Size of course affects the price of a shell, especially in the case of extremely large examples or immature shells. Thus a good price list should state *approximate* size and inform you if the specimen is immature. Immature shells have weakly developed teeth, often have a different shape and color pattern, are very light in weight, and are more fragile than adults. Immatures of some species, such as *Cypraea tigris* and *C. aurantium*, may be as large as fully mature adults.

A specimen without at least locality data is, unless very rare, worthless. Dealers provide the most accurate information available to them, but admittedly this is sometimes subject to doubt. Where bad locality information is provided, it is almost always due to poor locality being provided to the dealer by his wholesaler, etc. You will soon develop a "fifth sense" as to whether or not locality data is accurate; avoid dealers who consistently provide bad information and usually misidentify their shells.

Correct identification of shells is always a problem, as seldom do two taxonomists or collectors agree on the proper name for species and varieties. Australian dealers and collectors, for example, use a very different nomenclature for the genera and species of some cowries than do collectors in the United States. It is a wise precaution to always double-check the identification of either purchased material or of material obtained through exchange.

Reputable dealers provide a reasonable guarantee of satisfaction with the material they sell and will cheerfully replace defective or misidentified shells—as long as the error is theirs and not yours. They send the best specimens available at the price and with the best identification possible, but don't expect perfection at bargain prices. You get what you pay for.

Many dealers regularly advertise in at least two magazines which every shell enthusiast should subscribe to. *Of Sea and Shore* magazine (P.O. Box 33, Port Gamble, Washington, USA 98364) is a quarterly containing news and articles of interest to all shell collectors from the beachcomber type to the professional collector. Each issue carries at least four pages of dealer ads from around the world. *Hawaiian Shell News* (Hawaiian Malacological Society, P.O. Box 10391, Honolulu, Hawaii, USA 96816) is a thin but highly informative monthly which will keep even advanced collectors up-to-date.

Additionally, the Italian shelling magazine *La Conchiglia* (Via C. Federici, 1—00147 Rome, Italy) has been running many articles of great interest to cowry collectors, including the descriptions of new species and subspecies. These are excellently illustrated in color and essential to any advanced collector.

SCHILDERORUM Iredale: Broadly oval; lateral calluses thick and extending onto dorsum; teeth fine, not extending onto base; dorsum brownish with darker bands; base mostly brown except for white area along aperture. See *C. sulcidentata, C. ventriculus, C. carneola*. Central Pacific Ocean. 21.9-43.3 (36.8) mm.

MINORIDENS Melvill: Cylindrical; tips of aperture reddish; base not spotted; dorsum with many small but well-marked brown spots; middorsal band nearly solid. See *C. fimbriata, C. serrulifera, C. hammondae.* Western Pacific Ocean. 9.4-12.2 (10.1) mm.

Sizes indicated are the usual range, not necessarily the extremes; sizes in parentheses are those of the illustrated specimen.

HAMMONDAE Iredale: Oval to cylindrical; sides and base distinctly spotted; brown dorsal bands four, interrupted, darker at edges. See *C. serrulifera, C. microdon, C. irrorata*. Northern Australia to Philippines. 10.6-17.2 (15.4) mm.

BECKII Gaskoin: Inflated-cylindrical; anterior teeth of inner lip fused; teeth of outer lip marked with brown; dorsum with large brown spots and smaller white ones. See *C. macandrewi, C. martini, C. punctata.* Pacific Ocean. 8.4-15.9 (14.1) mm.

SERRULIFERA Schilder & Schilder: Slender, tapering; tips of aperture pinkish to reddish; dorsum heavily spotted with small brown spots; four narrow brown dorsal bands about equally spaced; base not spotted. See *C. hammondae, C. microdon, C. minoridens*. Southeast-central Pacific Ocean. 8.7-12.8 (11.5) mm.

ZICZAC Linnaeus: A tan or brown shell with three broad bands of distinctive zigzag lines of white; base heavily spotted. See *C. diluculum*. Indo-Pacific. 12.3-23.9 (18.1) mm.

LUTEA Gmelin: Olive-brown above with two narrow and well marked white bands; dorsal spotting sparse, weak; base orangish, heavily spotted with darker. See *C. humphreysi.* Western Pacific Ocean and northwestern Australia. 9.4-22.0 (14.0) mm.

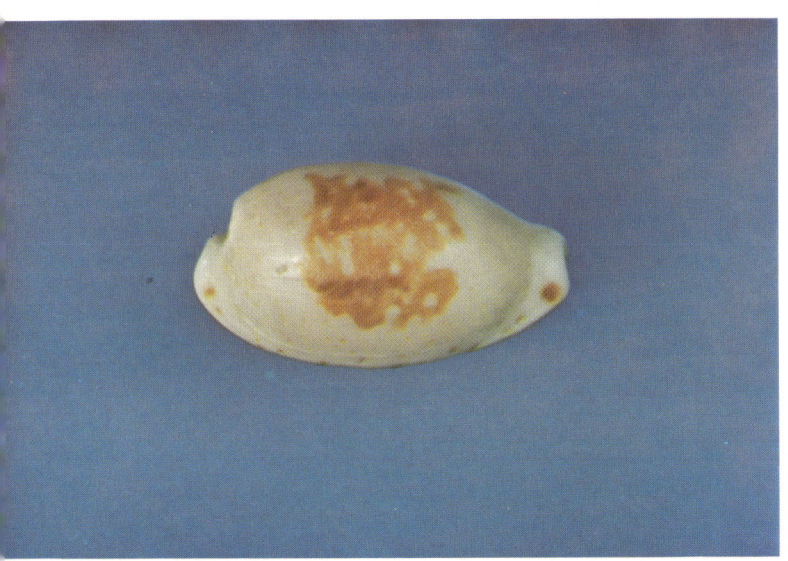

GOODALLII Sowerby: Small, the posterior end of inner lip produced; creamy, with large brown middorsal blotch and usually small basal spots and two spots near the tips; all spotting variable. See *C. saulae, C. contaminata, C. gracilis*. Central Pacific Ocean. 9.4-20.0 (13.3) mm.

ERYTHRAEENSIS Sowerby: Very similar to *C. stolida*, but the teeth often finer, often prolonged across the base. Red Sea and Gulf of Aden. 16.0-26.8 (17.4) mm.

PUNCTATA Linnaeus: Elongate-oval; tips moderately produced; teeth long, usually outlined with yellowish; dorsum white with scattered rounded dark brown to black spots of regular size; base with spots. See *C. irrorata, C. beckii, C. macandrewi.* Indo-Pacific. 9.4-22.0 (18.2) mm.

BARCLAYI Reeve: Teeth very heavy, stained with orange; base strongly convex; white, with numerous fine brown spots so closely spaced as to hide most of white background in fresh specimens; tips orange. Western Indian Ocean. 22-28 (25) mm.

CLANDESTINA Linnaeus: Oval; tips moderately prominent; teeth strong; dorsum cream, with very indistinct off-white bands and very fine brown zig-zag lines; base white. See *C. artuffeli*. Indo-Pacific. 10.8-25.0 (17.5) mm.

BISTRINOTATA Schilder & Schilder: A small, round, inflated shell with prominent tips; dorsum entirely granulate; three pairs of brown blotches (sometimes fused) along dorsal midline of shell; four brown basal spots. See *C. cicercula, C. globulus, C. mauiensis*. Indo-Pacific. 11.7-20.5 (20.2) mm.

IRRORATA Gray: Cylindrical; anterior teeth of inner lip heavy, pointed; tips white; dorsum brownish to bluish, with large, distinct dark spots, not the usual stippling of related species. See *C. hammondae, C. serrulifera.* Central Pacific Ocean. 9.4-14.4 (12.1) mm.

VREDENBURGI Schilder: Oval, inflated; the tips not very prominent, marked with four brown spots; anterior part of inner lip with a distinct depression within and a second row of teeth; dorsum heavily mottled with small brown spots and blotches, sometimes with a darker central blotch; three prominent dark brown bands dorsally; lateral spotting not extending onto base. See C. pallida, C. xanthodon. Indonesia. 17.9-26.4 (20.3) mm.

BICOLOR Gaskoin: Oval; margins and tips well developed; dorsum tan with two narrow white bands; a uniform fine brown reticulated net often present over dorsum; dark spots laterally and sometimes dorsally. See *C. piperita, C. angustata, C. declivis.* Southern Australia. 16.2-25.8 (22.8) mm. Was called *piperita* in first edition.

URSELLUS Gmelin: Elongate-oval; small; tips marked with four dark spots; teeth of inner lip long, extending across base; dorsal pattern of contrasting blue-gray bands on cream background. See *C. hirundo, C. kieneri*. Western Pacific Ocean. 9.4-12.6 (10.5) mm.

MACANDREWI Sowerby: Cylindrical; teeth relatively coarse, marked with brown on outer lip; anterior teeth on inner lip not fused; dorsum with large brown spots and scattered smaller white ones. See *C. punctata, C. beckii.* Red Sea. 11.0-23.2 (13.1) mm.

SAULAE Gaskoin: Small, the posterior end of inner lip produced; bluish to creamy tan, usually with a large brown middorsal blotch and distinct brown lateral spots; blotch may be absent or replaced by many small spots; teeth sometimes stained orange. See *C. contaminata, C. goodalli.* Western Pacific Ocean. 18.8-28.6 (20.6) mm.

HIRUNDO Linnaeus: Elongate to cylindrical; tips marked with four dark brown spots; teeth of inner lip fine, long, crossing the base. Dorsum variable in color, often bluish, with fine brown spots or a larger central blotch, seldom a conspicuous banded pattern. See *C. kieneri, C. ursellus, C. cylindrica.* Indo-Pacific. 13.2-22.4 (22.4) mm.

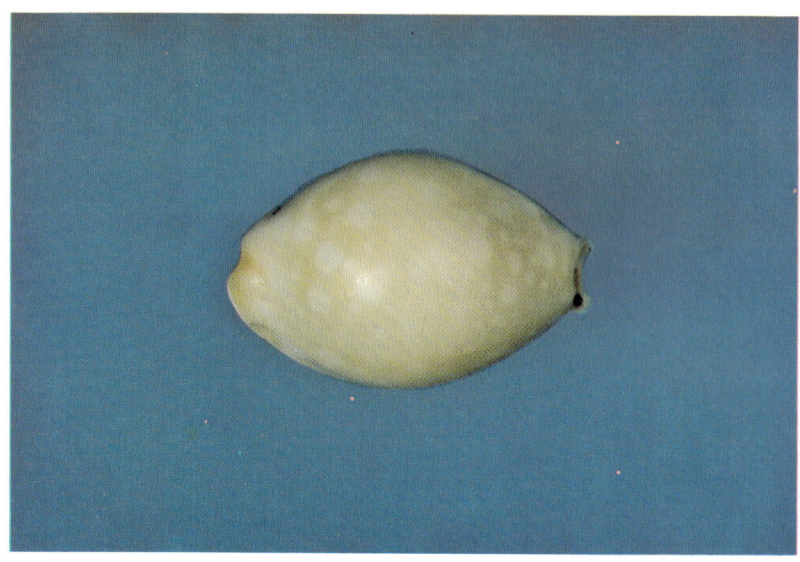

DILLWYNI Schilder: A small, oval shell; off-white to tan, with numerous large and small white spots over dorsum; indistinct brown blotches at tips of shell. Polynesia. 10.2-15.4 (13.0) mm.

SANGUINOLENTA Gmelin: Elongate-oval; posterior tip of inner lip poorly developed; teeth weak, especially those of inner lip; in life violet-tinged, but rapidly fading to brown and cream; spots present on base; dorsal pattern of brown mottling and a brown central blotch or band. See *C. zonaria*. Cape Verde Is. and Bulge of Africa. 19.2-26.6 (21.6) mm.

GLOBULUS Linnaeus: A small, round, inflated shell with prominent tips; dorsum smooth, even under low magnification; dorsum with small spots, no large blotches on spire or along middorsal line; base with or without spots. See *C. mauiensis, C. cicercula, C. bistrinotata.* Indo-Pacific. 9.8-23.6 (21.0) mm.

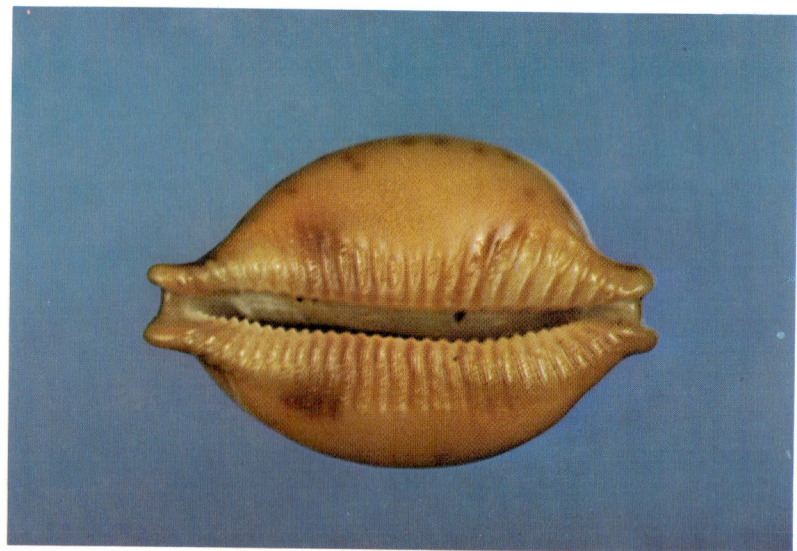

HIRASEI Roberts: Oval; tips produced (posterior tip of inner lip very short compared to that of outer lip); teeth weak, numerous; base solid white; dorsum cream with a large dark brown central blotch of variable size and shape; sides heavily spotted with small brown to purplish spots. See *C. langfordi, C. midwayensis, C. surinamensis.* Japan. 40.1-61.4 (55.2) mm.

OWENII Sowerby: Oval; teeth heavy, long, extending across base; tips marked with four dark spots; dorsum with fine spotting which extends onto base and nearly to teeth. Western Indian Ocean. 8.0-21.8 (14.3) mm.

INTERRUPTA Gray: Relatively elongate; teeth heavy, numerous, those on inner lip long; dorsum with four often interrupted brown bands on a mottled background, the two middle bands very close together and sometimes fused. See *C. pallidula, C. luchuana, C. summersi*. Ceylon to Java and the Philippines. 15.9-26.0 (20.5) mm.

Shells are often found beach worn beyond identification. This badly worn and decorticated (top layer of nacre and pigment missing) little shell would not be suitable for the collector. Unless very rare, such shells have no value.

MICRODON Gray: Elongate-cylindrical, the tips projecting in adults; teeth very fine; dorsum with two pale bands beaded with small brown spots at margins; base unspotted. See *C. hammondae, C. serrulifera.* Indo-Pacific. 11.4-14.7 (13.2) mm.

HUMPHREYSI Gray: Tan with three wide whitish bands dorsally, heavily covered with brown spots and blotches so as to obscure the ground color; base orange, with large dark spots. See *C. lutea*. Coral Sea to the Solomons and Fiji. 12.6-21.5 (19.1) mm.

CONTAMINATA Sowerby: Very small; posterior end of inner lip produced; creamy to pale purple, usually with indications of three broad brown bands; a dark brownish middorsal blotch or many small brownish spots may or may not be present; numerous basal spots; ends of aperture not stained brownish internally. See *C. gracilis, C. goodallii*. Indian Ocean and Western Pacific Ocean. 10.2-13.7 (13.1) mm.

This specimen of *C. macandrewi* was mistakenly labeled as *thomasi* in the first edition due to confusion of type specimens. The name *thomasi* is now applied to the species formerly called *ostergaardi*.

SUMMERSI Schilder: Oval-elongate; teeth heavy, those on inner lip long; teeth relatively coarse and few; dorsal pattern obscure, broad bands only vaguely indicated; anterior tip with brown smudges. See *C. pallidula, C. luchuana, C. interrupta.* Fiji area. 13.8-18.3 (14.1) mm.

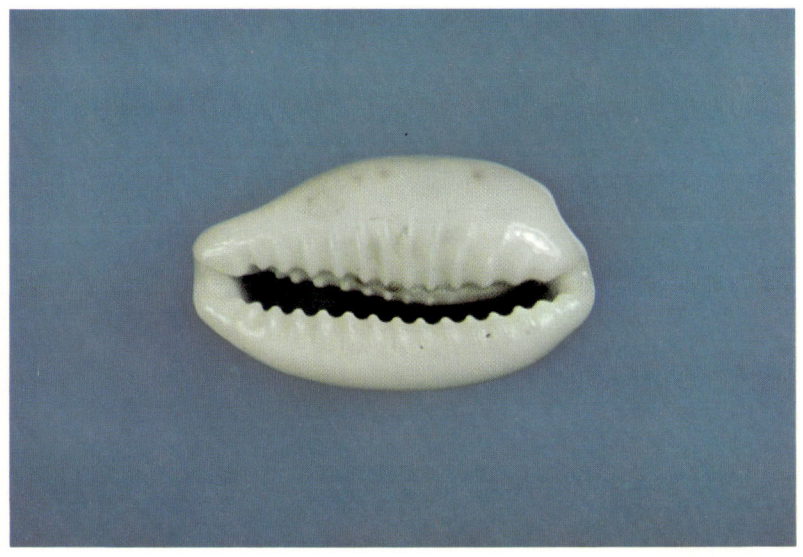

THOMASI Crosse: Oval, the margins and tips moderately produced; margins, base and tips white; dorsum cream with scattered large brown spots. See *C. gangranosa, C. helvola, C. spurca.* Hawaiian Islands. 9.5-24.2 (11.7) mm. Was called *ostergaardi* in first edition.

ISABELLAMEXICANA Stearns (subspecies of *isabella*): Very much like typical *isabella* but the tips more produced, brighter orange; sides of base (in adults) with a brownish ring. Pacific Central America. 31.4-49.3 (49.3) mm.

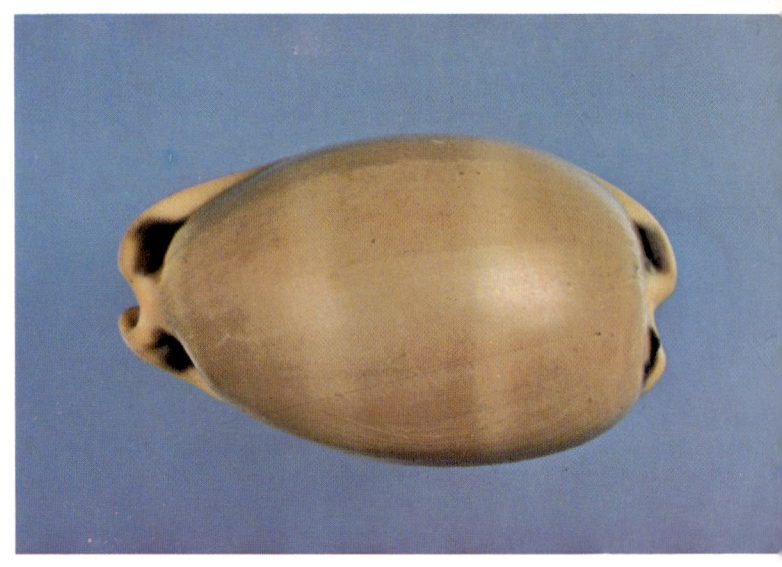

LURIDA Linnaeus: Oval-elongate; teeth not marked with brown; tips with four distinct nearly black spots; dorsum dark brown, fading after death to expose light bands; base tinged with brown. See *C. cinerea, C. pulchra*. Mediterranean and West Africa. 28.4-59.3 (39.5) mm.

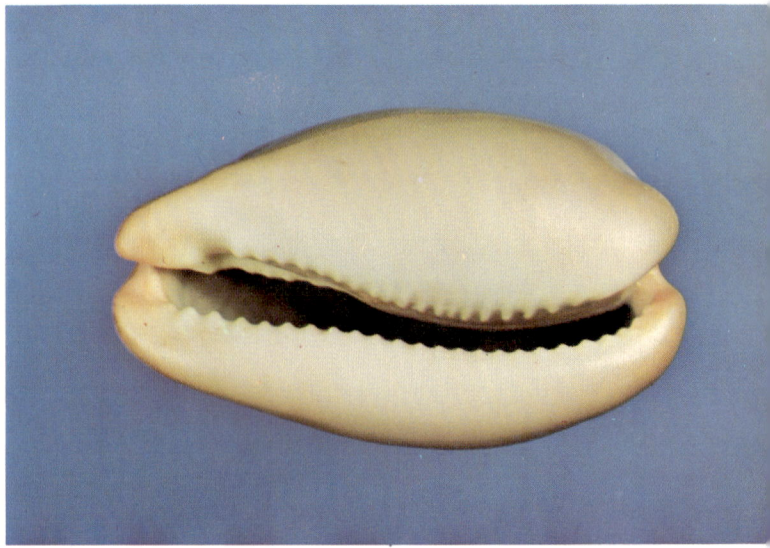

CAURICA Linnaeus: Elongate to oval, sometimes nearly circular; teeth usually heavy, interspaces yellowish to tan; margins usually well developed, sometimes irregular; margins with a row of regular blackish to purple spots, widely spaced; dorsum heavily mottled with fine brown spots and blotches, sometimes with a darker central blotch. Very variable. See *C. chinensis*. Indo-Pacific. 25.3-50.8 (40.8) mm.

FUSCORUBRA Shaw: Outer lip strongly projecting, inner truncate teeth present on both lips, numerous; dorsum with large brown blotches. See *C. cruickshanki, C. algoensis.* South Africa. 39.0-45.7 (43.2) mm.

CINEREA Gmelin: Oval; teeth strong, those of inner lip long and sometimes faintly marked with brown; dorsum deep brown fading to tan after death to expose light dorsal bands; dorsum and sides with fine black sparse to abundant specks; tips with four vague spots. See *C. lurida*. Western Atlantic Ocean. 20.6-40.0 (25.0) mm.

ASELLUS Linnaeus: Elongate; teeth heavy; a solid white shell with three broad nearly black dorsal bands of uniform width. Indo-Pacific. 14.3-22.5 (17.6) mm.

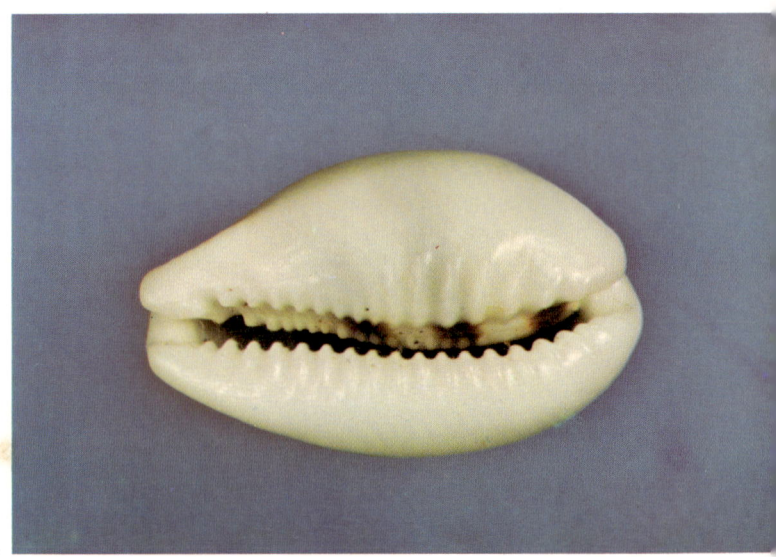

FULTONI Sowerby: Margins of adult shell depressed, irregular; teeth well developed, sometimes lightly colored; base with large brown spots; irregular brown dorsal figure. See *C. mus, C. teulerei.* South Africa. 50.1-65.8 (60.8) mm.

ERRONES Linnaeus: Elongate-oval; tips moderately prominent teeth short, relatively weak, especially on inner lip; teeth usuall white, seldom yellowish; base and tips white; dorsum greenish tan heavily mottled with small brown spots and blotches; usuall a large dark brown irregular central blotch and vaguer blotche over anterior and posterior ends of dorsum. See *C. ovum*. Wester Pacific Ocean and eastern Indian Ocean. 16.7-42.3 (37.1) mm.

SULCIDENTATA Gray: Broadly oval; lateral calluses thick; teeth heavy, extending well onto base; dorsum brownish with darker bands, much brighter in life but rapidly fading; base tan. See *C. schilderorum*. Hawaiian Islands only. 23.4-67.2 (34.5) mm.

ONYX Linnaeus: Shell oval, tapering anteriorly; aperture strongly curved posteriorly; coloration variable, but base and sides deep brown to blackish; dorsum solid deep brown, deep brown with narrow lighter bands, or nacreous white; one variation is almost all white except for vague brown dorsal bands. Indo-Pacific. 24.2-46.0 (37.4) mm.

EXUSTA Sowerby: Very similar to *C. talpa*, but the teeth finer. Red Sea and Gulf of Aden. 63.7-72.3 (63.7) mm.

ARMENIACA Verco: Elongate; tips extremely well developed and long; aperture very strongly curved posteriorly; teeth of inner lip weak at center; spire area large, prominent, depressed; base whitish or orangish; dorsum white to cream with tan mottling of variable extent. See *C. marginata, C. friendii.* Southern Australia. 66.0-116.4 (93.5) mm. Includes *C. hesitata* of first edition.

SUBTERES Weinkauff: Elongate to cylindrical; posterior tip greatly produced; outer lip thin, sharp marginally; large brown marginal spots present; dorsum brownish with three indistinct darker bands. See *C. teres.* Eastern Polynesia. 14.2-27.2 (27.2) mm.

ROSSELLI Cotton: Oval, adult shells very wide at posterior third; dorsum greatly humped; aperture strongly curved posteriorly; shell almost entirely black above and below (really a deep dark brown) except for a whitish tan dorsal area in some shells. See *C. decipiens*. Western Australia. 46.0-64.4 (46.0) mm.

LYNX Linnaeus: Elongate to oval; tips moderately prominent; teeth stained with bright orange; dorsum tan to bluish, mottled; large blackish spots distributed over entire dorsal area and on margins. Indo-Pacific. 19.8-65.2 (41.9) mm.

PULCHRA Gray: Oval-elongate; teeth heavily marked with brown, those of inner lip continued onto base at middle; tips with four nearly black spots; dorsum deep brown when fresh, fading to expose two light bands after death. See *C. lurida*. Red Sea to Gulf of Oman. 25.5-76.3 (49.2) mm.

ARGUS Linnaeus: Cylindrical; large; teeth outlined with brown; shell cream, dorsally with traces of three broad darker bands; dorsum with many brown circles of various sizes and darkness; base with two or four rectangular black blotches near mouth. Indo-Pacific. 63.0-93.0 (77.0) mm.

TURDUS Lamarck: Shell inflated; teeth of inner lip often few, relatively weak; white with numerous small, compact spots dorsally, becoming larger and darker toward margins; margins usually a raised callus. See *C. tigris*. Western Indian Ocean. 19.7-47.4 (36.5) mm.

PYRUM Gmelin: Oval, tapered anteriorly; teeth large, whitish; dorsum brownish with darker clouding and blotching; base and tips usually orange to yellowish, occasionally whitish. See *C. zonaria*. Mediterranean and West Africa. 25.2-48.9 (39.8) mm.

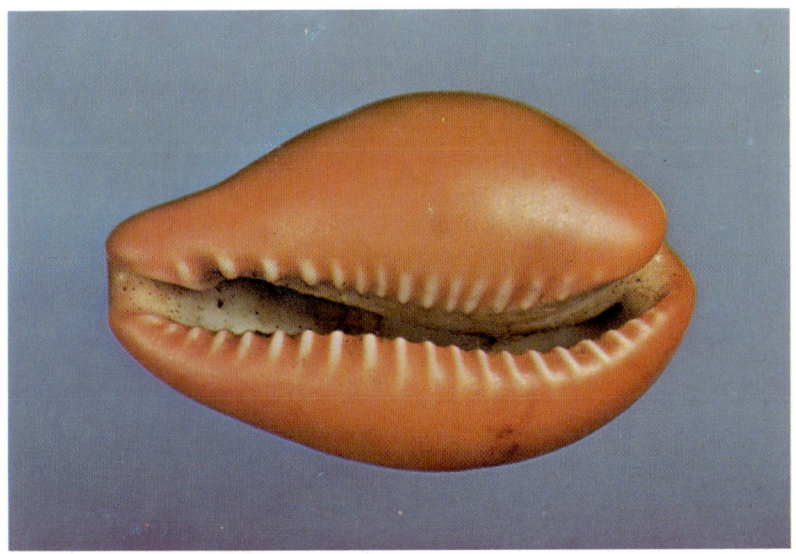

MUS Linnaeus: Margins of adult shell depressed, irregular; teeth well-developed, stained with brown; base brownish, not spotted; black spots concentrated along middorsal area. See *C. teulerei, C. fultoni.* Atlantic coast of Colombia and Venezuela. 31.6-57.6 (41.1) mm.

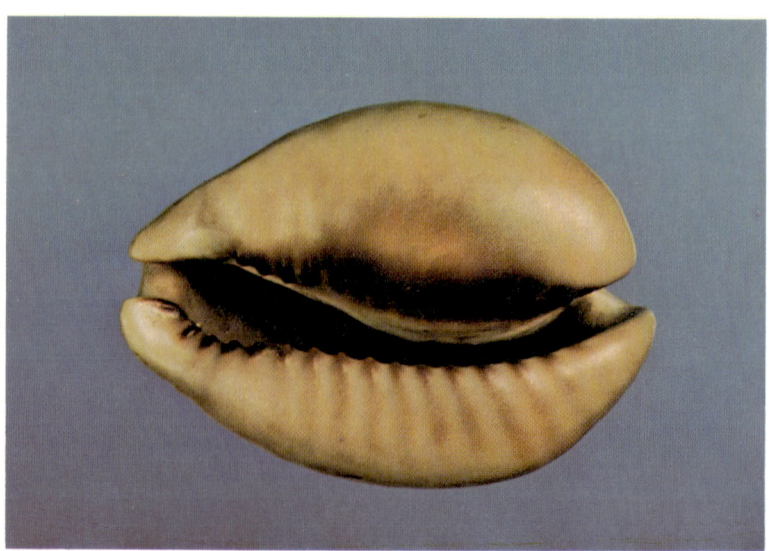

AURANTIUM Gmelin: Oval, greatly inflated; teeth marked with orange; dorsum solid golden orange to nearly tan, sometimes with vague lighter bands or spots visible; base usually solid white. Unmistakable. See *C. carneola*. Pacific Ocean. 58.0-115.1 (115.1) mm.

DAYRITIANA Cate: Oval, inflated; tips prominent; teeth few, very long and coarse; spire with a dark blotch; base and tips white, dorsum greenish tan, heavily mottled with fine brown dots and blotches. See *C. ovum, C. luchuana, C. pallidula.* Philippine Islands. 15.5-21.5 (17.0) mm.

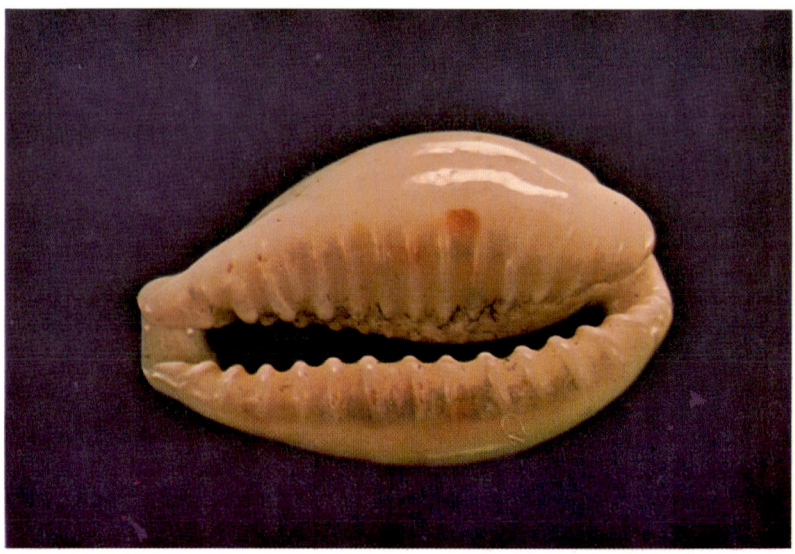

ARTUFFELI Jousseaume: Elongate-oval; tips moderately produced; dorsum deep brown with a darker central band; zig-zag brown lines usually absent. Very similar to *C. clandestina*. Japan and Ryukyus. 16.2-22.6 (16.2) mm.

VENUSTA Sowerby: Variable; usually broadly oval to nearly circular; tips well developed; aperture curved posteriorly; teeth absent over most of inner lip; base commonly white or cream with reddish-brown lateral spots; dorsum cream with varying mottling of brown to reddish brown, sometimes nearly all dark brown. See *C. friendii, C. marginata.* Southwestern Australia. 54.6-80.6 (76.5) mm.

LEVIATHAN Schilder & Schilder (variety of *carneola*): Very large (over 50mm) shells usually with nodules in the marginal callus and the spire heavily nacred over in adults. The radula is quite different from that of typical *carneola* but this is a doubtful character not usable by the collector. Best developed in the Polynesian area, but similar shells may occur throughout the Indo-Pacific. 53.4-110.2 (73.2) mm.

VENTRICULUS Lamarck: Oval, depressed, the sides nearly parallel; lateral calluses very heavy, extending in adult shells to cover nearly the entire dorsum; teeth fine, short; brown with darker bands in young, the pattern obscured in adults; margins with fine white wrinkles. See *C. schilderorum*. Southern Pacific Ocean. 41.6-66.0 (58.2) mm.

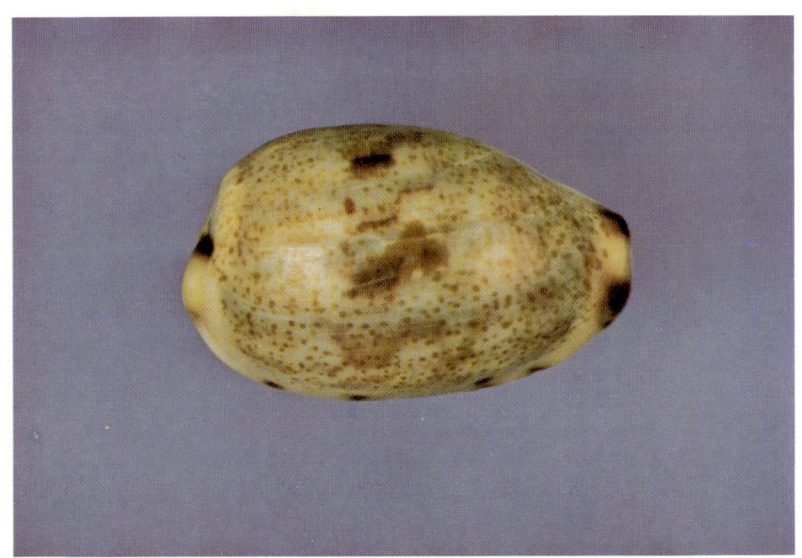

GRACILIS Gaskoin: Oval, slightly tapering; small; aperture stained with brown at both ends; a dark brown middorsal blotch and dark basal spots; tips brown-spotted. See *C. saulae, C. contaminata, C. fimbriata*. Indian Ocean and Western Pacific Ocean. 12.4-26.3 (16.0) mm.

MAPPA Linnaeus: Oval, inflated; background of dorsum and base cream or tan, the dorsum overlaid by a dense network of longitudinal brown lines and scattered brown circles; mantle mark a broad, longitudinal undulating light line, very irregular in outline. See *C. arabica*. Indo-Pacific. 47.5-90.0 (79.6) mm.

VALENTIA Perry: Large, heavily inflated, oval; tips prominent, folded; teeth very fine, numerous; dorsum pale tan with scattered spots of darker brown and two lateral irregular brown blotches; base white. Eastern Indian Ocean to Fiji in small colonies. 63.5-97.8 (89.6) mm.

REEVEI Sowerby: Oval, greatly inflated; very thin and juvenile in appearance; dorsum tan with darker bands; base white, tips rose-colored. See *C. carneola*. Western Australia to South Australia. 33.5-38.0 (35.1) mm.

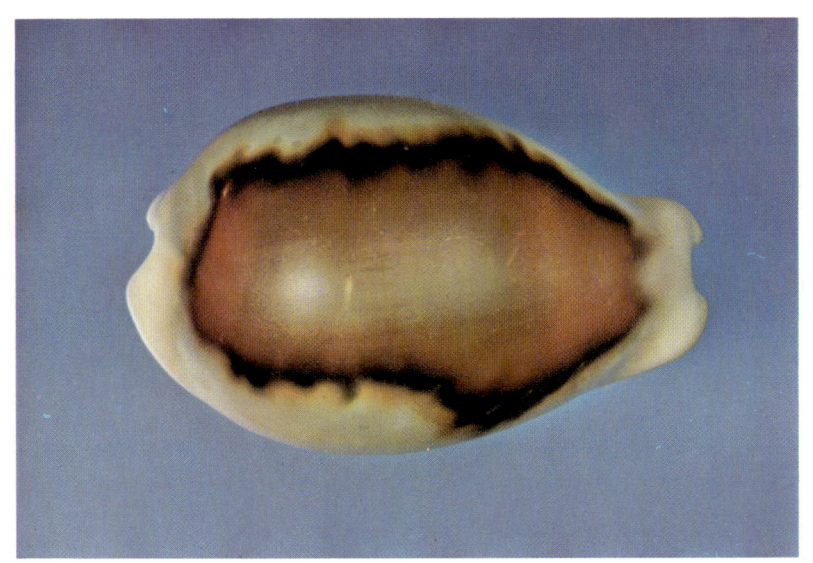

SPADICEA Swainson: A whitish shell with a distinctive deep brown dorsal blotch covering most of the dorsum and often bordered by darker. California. 35.8-59.8 (45.4) mm.

EBURNEA Barnes (subspecies of *miliaris*): Dorsum completely white, lacking any pattern (occasional shells have a faint yellowish or grayish sheen). Melanesia. 32.4-48.6 (41.2) mm.

CHILDRENI Gray: Oval, the tips projecting; teeth prominent, forming long projections ("fins") which extend entirely around shell to midline of dorsum; yellowish. Indo-Pacific. 11.8-22.3 (21.9) mm.

BOIVINII Kiener: Elongate, somewhat cylindrical; teeth heavy; dorsum slate to milky with bluish tones; base white; dorsum with numerous small dark (often bluish) spots and a brown blotch over the spire; often a brown mantle line from tip to tip. Indo-Pacific. 16.8-36.4 (25.5) mm.

KIENERI Hidalgo: Elongate to cylindrical; tips marked with four dark spots; teeth of inner lip short, not extending across base; dorsal pattern of blue-gray bands on cream, not always distinct; fine dorsal mottling and larger central figure also present. See *C. ursellus, C. hirundo.* Indo-Pacific. 10.1-22.2 (20.2) mm.

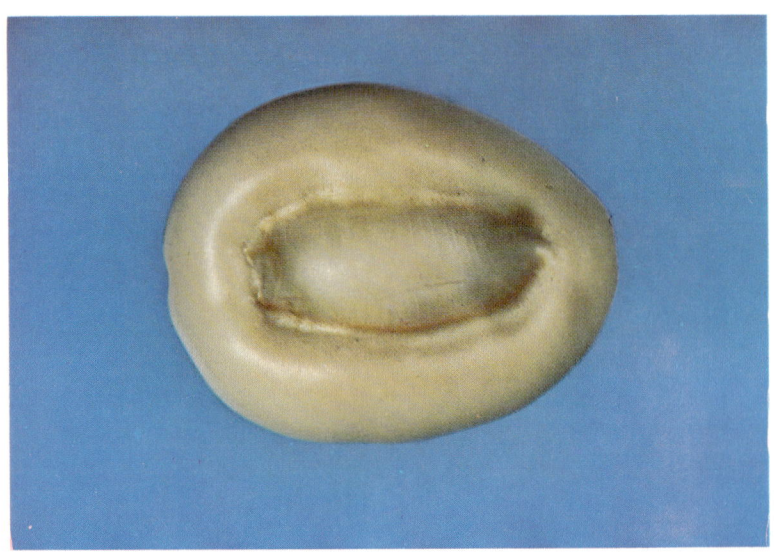

OBVELATA Linnaeus (subspecies of *annulus*): Oval; margins of shell greatly infalted in adults, with a distinct circular depression separating margin and middorsal area. Eastern Polynesia. 14.1-26.5 (20.4) mm.

Photos courtesy Drs. Rehder & Rosewater, National Museum of Natural History.

KINGAE Rehder & Wilson: Inflated, with strong calluses; teeth fine, the anterior columellar teeth connected by a ridge; orangish with white spots; base white; usually white nacre spots on dorsum. See *C. bernardi, C. cernica, C. englerti.* Pitcairn Island. 14-19mm.

ALBUGINOSA Gray: Oval, the margins and tips prominent and produced in adult shells; dorsum brownish, heavily tinged with purplish, as is the base; numerous small white spots on the dorsum, each outlined with a narrow brown ring, producing an "out of focus" effect; basal spots weak or absent. See *C. poraria, C. marginalis*. Eastern Pacific Ocean. 18.9-28.8 (28.8) mm.

PIPERITA Gray: Oval; margins and tips well developed; color very variable; usually the marginal spots are faint or absent, not heavily developed on the base; dorsum with two narrow brown bands closely placed just off-center of dorsum; no dorsal spotting; shells may be entirely white or entirely dark. See *C. angustata, C. bicolor, C. pulicaria, C. declivis.* Southern Australia. 20.7-27.0 (27.0) mm. Was called *comptoni* in first edition.

ANNULUS Linnaeus: Oval; teeth heavy, coarse; base concave; creamy brown with a bright orange circle dorsally. See *C. moneta, C. obvelata*. Indo-Pacific. 12.6-29.0 (24.2) mm.

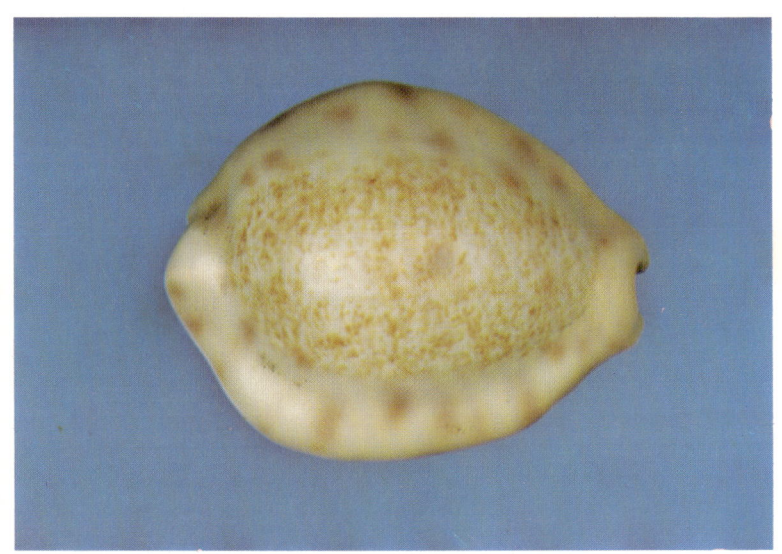

COLOBA Melvill (subspecies of *chinensis*): Margins greatly swollen, often nodulose; base orange; clear round spots on dorsum reduced or absent. Northern Indian Ocean.

OVUM Gmelin: Oval, inflated; tips prominent; teeth short, coarse, not extending over base; teeth stained yellowish to orange-red; base and tips white; dorsum greenish tan heavily mottled with fine brown spots and blotches, sometimes with large brown central blotch. See *C. dayritiana, C. errones.* Western Pacific Ocean. 22.5-38.9 (27.2) mm.

LENTIGINOSA Gray: Elongate-oval; tips prominent; teeth relatively few, heavy; dorsum and base cream, heavily spotted with nearly uniform brown spots except near the aperture. See *C. owenii*. Red Sea to southern India. 20.5-32.3 (28.2) mm.

GRANULATA Pease: A rough cowry heavily studded with nodules; oval, depressed, tips not prominent; surface dull; orange and violet to grayish. See *C. nucleus.* Hawaiian Islands and southeastern Polynesia. 14.9-43.0 (24.3) mm.

ANGUSTATA Gmelin: Oval; margins and tips well developed; dorsum uniform brown without distinct spotting or banding; margins and often base with large dark brown spots, often densely developed. See *C. bicolor, C. piperita, C. declivis*. New South Wales to South Australia. 22.3-35.8 (28.3) mm.

CARNEOLA Linnaeus: Cylindrical, inflated, very light in weight; teeth marked with purple; dorsum tan to orangish with evenly spaced darker bands; the lateral calluses are smooth, not nodulose; specimens rarely exceed 3 inches. See *C. leviathan, C. schilderorum*. Indo-Pacific. 21.0-89.0 (64.6) mm.

CUMINGII Sowerby: Dorsal light spots large or small, very distinct and often outlined with brown; lateral spots distinct, dark brown; elongate to rather oval, usually tapering at both ends. See *C. cribraria, C. gaskoini, C. catholicorum.* Polynesia. 10.6-27.8 (27.8) mm.

MARGARITA Dillwyn: Small, humped; tips prominent, straight or twisted; teeth at middle of columella usually short; smooth dorsally, granules not developed; whitish to pale tan or yellowish with small brown spots; spire spot absent or very small. See *C. cicercula, C. mauiensis*. Indo-Pacific. 10-20mm.

DECLIVIS Sowerby: Oval; margins and tips poorly developed; dorsum tan with numerous small darker brown spots, often most visible laterally; juvenile shells may be banded. See *C. bicolor, C. angustata, C. piperita.* 19.2-27.6 (27.6) mm.

Hybrids between C. capensis and C. edentula have been described as a full species, C. amphithales. Such hybrids have some characters of both species, usually with the dorsum smooth and spotted like edentula and the base with teeth and ridges like capensis. South Africa. 25.4-29.7 (25.9) mm.

QUADRIMACULATA Gray: Cylindrical; teeth heavy, those on base very long; tips with four black blotches; dorsum mottled with fine brown dots on a greenish tan background. See *C. pallidula, C. luchuana*. Western Pacific Ocean and northwestern Australia. 17.4-32.0 (24.2) mm.

EDENTULA Gray: Inflated; outer lip projecting, inner truncate; teeth absent or nearly so, occasionally present on outer lip; brownish spots and mottling on cream, often a darker central figure dorsally. See *C. algoensis*. South Africa. 20.4-29.4 (28.4) mm.

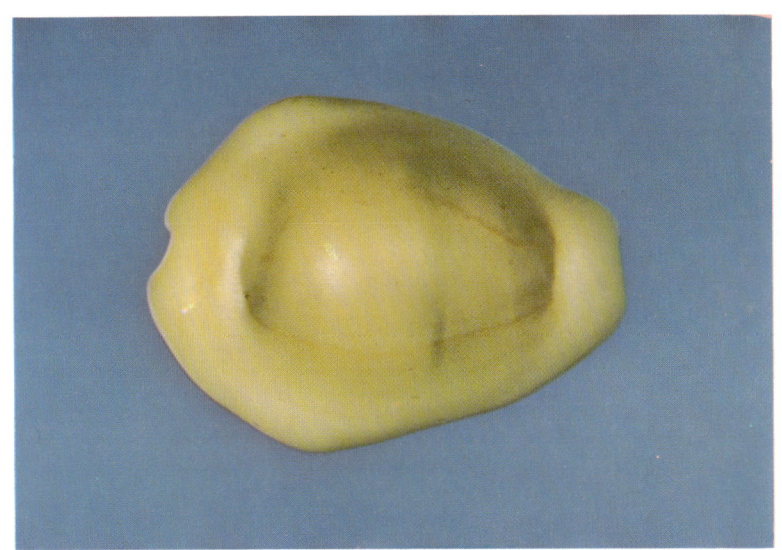

MONETA Linnaeus: Margins of shell generally widened near posterior third; teeth relatively weak, especially on inner lip; coloration and shape highly variable. See *C. annulus, C. obvelata.* Indo-Pacific. 12.2-38.5 (23.6) mm.

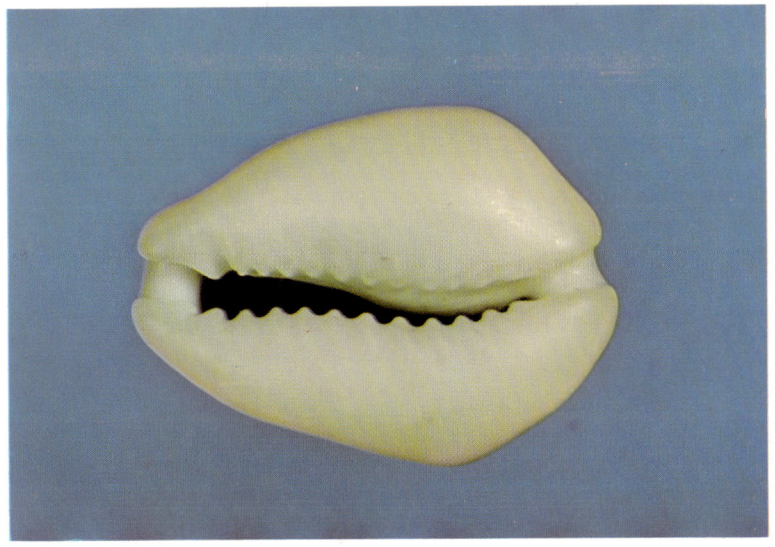

WALKERI Sowerby: Oval-elongate; teeth very fine posteriorly, large anteriorly; teeth and interior stained purplish; dorsum tan to dark brown with two narrow cream bands equally dividing the dorsum; base brown to white; no small white spots on base. See *C. bregeriana*. Indian Ocean, Java, northern Australia, Philippines. 18.0-37.0 (30.0) mm.

LABROLINEATA Gaskoin: Elongate-oval; margins and tips very prominent; aperture not reddish; tips above with dark blotches; dorsum with many white spots and no dark spots; base lightly spotted with brown. See *C. gangranosa, C. lamarckii*. Pacific Ocean. 13.6-25.5 (25.5) mm.

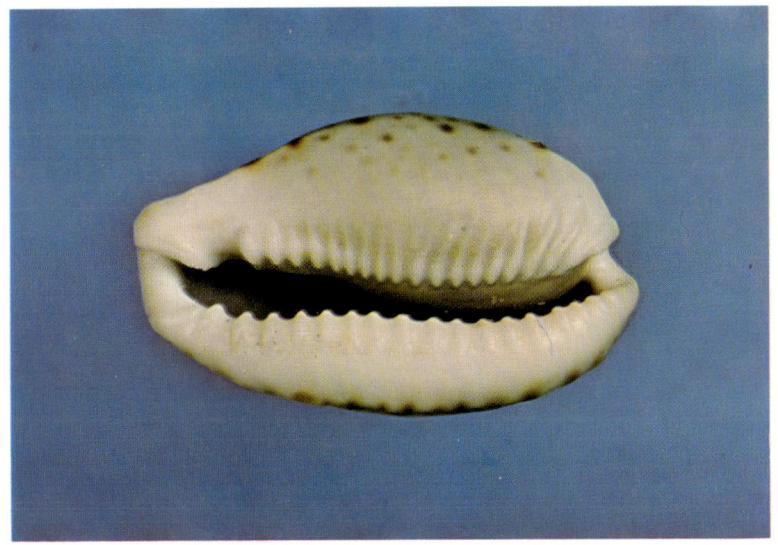

CYLINDRICA Born: Elongate, cylindrical, sides nearly parallel; tips marked with four dark spots; teeth of outer lip coarse; anterior teeth of both lips long, well developed; dorsum variably mottled with brown, usually with a larger central blotch. See *C. hirundo, C. kieneri, C. caurica.* Western Pacific Ocean and northwestern Australia. 19.4-41.4 (26.1) mm.

COXENI Cox.: Cylindrical; teeth heavy, those on base very long; base, margins, and tips white; dorsum with large brown blotches in an irregular pattern, the blotches variable in size and extent. Solomons to New Guinea. 16.2-25.4 (24.1) mm.

ROBERTSI Hidalgo: Oval; margins prominent; base strongly convex; teeth relatively coarse; dorsum with heavy brownish mottling, not forming distinct spots on the margins. See *C. arabicula*. Eastern Pacific Ocean. 17.2-30.9 (26.9) mm.

LIMACINA Lamarck: Dorsum of adults with distinct nodules, especially at the sides; relatively elongate shape; teeth of inner lip all about equal in length, not over half width of base at middle; blackish to bluish gray with small white spots. See *C. staphylaea*. Indo-Pacific. 19.6-35.7 (26.9) mm.

ALGOENSIS Gray: Inflated, outer lip projecting, inner truncate; teeth present on both lips, weak; brown spots and mottling on cream. See *C. edentula*. South Africa. 18.6-27.5 (24.8) mm.

STOLIDA Linnaeus: Oval; teeth large, coarse, somewhat prolonged; bluish cream to pale tan with a large deep brown middorsal blotch and usually similar markings on the sides; tips with four brown spots, often indistinct. See *C. erythraeensis*. Indo-Pacific. 16.0-36.7 (27.5) mm.

PETITIANA Crosse (variety of *pyrum*): Somewhat smaller than typical *pyrum*, with the base tan to creamy orange. Deeper water off West Africa. 21.9-25.8 (21.9) mm.

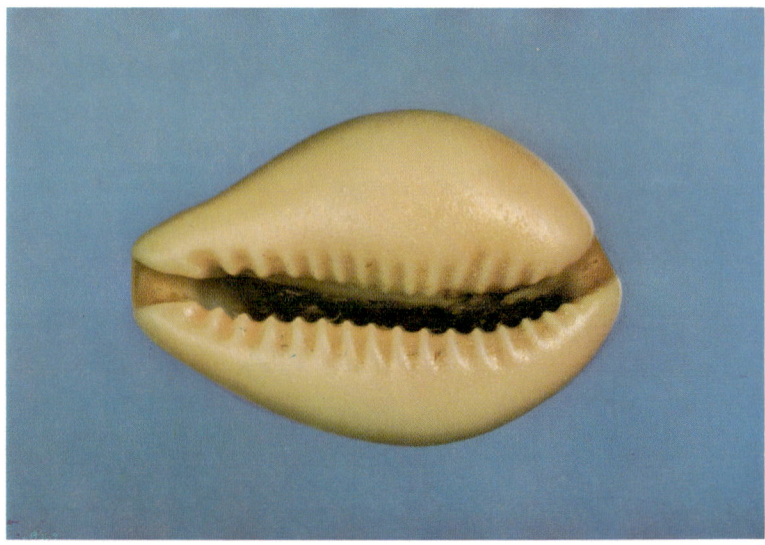

GANGRANOSA Dillwyn: Elongate-oval; margins and tips very prominent; anterior and posterior apertures stained reddish; tips above with four deep brown blotches; dorsum with many white spots on a tan background and a few larger dark spots. See *C. labrolineata, C. thomasi.* Indian Ocean and western Pacific. 12.3-23.1 (22.5) mm.

NEBRITES Melvill: Shell oval; teeth heavy, long, lined with brown; midsides of shell with two dark blotches, these not extending onto base; basal spotting heavy, dark brown. See *C. erosa*. Northwestern Indian Ocean. 23.3-34.6 (28.7) mm.

HISTRIO Gmelin: Cylindrical, inflated; base convex; teeth marked with brown; small but distinct blotch on spire; dorsum with large open reticulations and only a trace of longitudinal lines. See *C. eglantina, C. grayana*. Indian Ocean. 23.0-73.0 (68.2) mm.

PULCHELLA Swainson: Teeth prominent, darkly stained with black, extending across most of base; large dark brown middorsal blotch and lateral blotches usually present; tips dark. See *C. pyriformis, C. subviridis.* Northern Indian Ocean and western Pacific Ocean. 24.8-45.4 (38.6) mm.

SUBVIRIDIS Reeve: Teeth prominent, short, not stained with darker; base without spots, whitish; dorsum with usually large dark brown middorsal blotch and lateral blotches. See *C. pyriformis, C. pallida, C. xanthodon.* New Guinea and Australia; New Caledonia. 23.7-38.2 (32.5) mm.

TESTUDINARIA Linnaeus: Cylindrical; large; teeth fine; cream background color visible middorsally, but otherwise obscured by heavy black or dark brown mottling and spotting; base tan; dorsum heavily sprinkled with small white dots embedded in the nacre. Indo-Pacific. 79.5-130.8 (105.8) mm.

DECIPIENS Smith: Oval, greatly inflated and humped in adults; aperture nearly straight posteriorly; base and margins of shell deep brown, almost black; dorsum cream heavily mottled with deep brown, sometimes nearly all brown. See *C. rosselli*. Northwestern Australia. 49.4-59.2 (58.9) mm.

ARABICULA Lamarck: Oval; margins prominent, depressed; teeth relatively fine; dorsum mottled with brown lines; margins tinted with rose and bearing prominent black spots in a regular series; base and teeth white. See *C. robertsi, C. stercoraria, C. arabica* group. Eastern Pacific Ocean. 21.8-31.5 (31.5) mm.

XANTHODON Sowerby: Oval, inflated; tips prominent, marked with more-or-less distinct dark brown spots; inner lip without secondary teeth within anteriorly; spire blotch present; teeth stained yellowish, as usually is the base; dorsum heavily mottled with fine brown spots and blotches and with a pair of darker median bands. See *C. pallida, C. vredenburgi.* Eastern Australia. 21.4-35.5 (29.9) mm.

CERVINETTA Kiener: Inflated but relatively flat and cylindrical; anterior aperture wide; inner lip strongly projecting; interior purplish; teeth marked heavily with brown, coarse. See *C. cervus, C. zebra*. Eastern Pacific Ocean. 35.4-103.5 (80.9) mm.

PANTHERINA Solander: Inflated; tips produced; color variable, usually whitish with numerous small dark brown spots; sides and base white. See *C. tigris*. Red Sea and Aden. 57.5-76.5 (73.8) mm.

FUSCODENTATA Gray: Somewhat elongate, outer lip and inner lip both projecting; teeth of outer lip widely separated, those of inner weak; teeth prolonged into ridges ("fins") not extending dorsally, the ridges marked with brown; dorsum densely mottled with brown. See *C. capensis*. South Africa. 26.8-36.4 (31.9) mm.

PICTA Gray (subspecies of *zonaria*): Very much like *zonaria* but somewhat narrower and with finer and weaker teeth. Cape Verde Islands. 29.0-32.6 (30.5) mm.

NIVOSA Broderip: Brown with circular pale areas lacking pigment (not actually white); pale areas numerous, sometimes confluent; aperture curved posteriorly. See *C. vitellus, C. broderipii, C. camelopardalis.* Northern Indian Ocean. 47.4-62.0 (47.4) mm.

STERCORARIA Linnaeus: Heavy, inflated; lateral margins usually prominent; color variable, usually heavily mottled dorsally and on the margins; teeth often tinged with brown. See *C. arabicula, C. hesitata*. West Africa. 25.6-85.8 (57.2) mm.

PYRIFORMIS Gray: Teeth prominent, darkly stained with brown, not extending over much of base; dorsal blotches light brown, often vague, forming indistinct broad bands. See *C. pulchella, C. subviridis.* Northern Indian Ocean and western Pacific. 20.4-34.0 (33.6) mm.

SURINAMENSIS Perry: Elongate-oval; tips prominent; aperture strongly curved posteriorly; teeth heavy, stained with dark brown; dorsum and base orangish-brown, spotted or mottled dorsally with darker brown; spotting often fusing laterally to produce a nearly uniform dark brown band. See *C. teramachii, C. hirasei.* Northern South America to the Florida Keys. 32.6-38.7 (38.7) mm.

TIGRIS Linnaeus: Large, inflated shell; teeth of inner lip numerous; tips weakly produced; color very variable, from heavily spotted and mottled to nearly white; base white, with spots encroaching on sides. See *C. turdus, C. pantherina.* Indo-Pacific. 42.1-153 (89.5) mm.

MAURITIANA Linnaeus: Oval, inflated; very depressed, prominent margins; base and margins black; dorsum black with scattered tan spots. Indo-Pacific. 11.0-90.1 (90.1) mm.

CAPENSIS Gray: Inflated, outer lip projecting, inner somewhat extended; teeth produced into fine ridges ("fins") that extend across dorsum of the shell; brownish with darker mottling and a brown dorsal figure. See *C. edentula, C. fuscodentata.* South Africa. 26.8-35.5 (30.3) mm.

ZONARIA Gmelin: Oval, inflated; tips prominent; teeth strong; dorsum dull cream to tan, usually darker near the tips and with a broad zig-zag zone of dark brown at center; base spotted. See *C. sanguinolenta, C. picta*. West coast of Africa. 21.5-35.0 (35.0) mm.

GRAYANA Schilder (subspecies of *arabica*): Cylindrical, distinctly humped posteriorly; base convex; spire blotch absent or weakly indicated; tips relatively longer. See *C. arabica, C. histrio, C. eglantina.* Red Sea to Pakistan. 41.5-65.8 (64.0) mm.

DEPRESSA Gray: Broadly oval, the margins very prominently developed; base convex; teeth long and coarse, marked with brown; no dark blotch on inner lip; dorsum with an even, open reticulation leaving nearly circular uniform light spots; margins spotted. Indian Ocean; central Pacific Ocean. 25.0-55.5 (42.3) mm.

SPURCA Linnaeus: Elongate to oval; teeth heavy, those of anterior outer lip and posterior inner lip greatly prolonged; margins and tips prominent, the margins usually with a broken or complete row of small brown spots; dorsum brownish to yellowish orange with white spots or mottling, the spots often poorly defined. See *C. cernica, C. thomasi*. Atlantic Ocean. 20-35 mm.

DILUCULUM Reeve: A nearly black shell with two or three bands of distinctive zigzig white lines; base usually spotted. See *C. ziczac*. Western Indian Ocean. 13.6-32.0 (30.7) mm.

ARABICA Linnaeus: Subcylindrical; base flat or concave; teeth marked with brown; no dark blotch on spire; dorsal pattern of small reticulations over longitudinal lines. See *C. eglantina, C. grayana*. Indo-Pacific. 27.7-81.5 (66.6) mm.

CERVUS Linnaeus: Greatly inflated; inner lip only weakly projecting; anterior aperture wide; interior purplish; teeth marked with brown, relatively numerous; dorsal spots not ocellated. See *C. cervinetta, C. zebra.* Florida, Bahamas, eastern Mexico. 95.0-151.4 (119.5) mm.

CHINENSIS Gmelin: Oval to somewhat elongated; margins well developed; tips prominent; teeth heavy, especially those of outer lip, the interstices stained with yellow or orange; dorsum tan, reticulated with cream, often indistinctly blotched; large to small marginal spots present in regular series, overlapping onto base. See *C. caurica*. Indo-Pacific. 13.1-48.6 (29.9) mm.

PALLIDA Gray: Oval, inflated; tips not very prominent; inner lip without a depression and second row of teeth on its inner anterior margin; base and teeth white; dorsum heavily mottled with fine brown spots and blotches, often with a dark brown central blotch; lateral spotting not extending onto base. See *C. xanthodon, C. vredenburgi.* Northern Indian Ocean to Borneo. 24.3-30.5 (30.5) mm.

MACULIFERA Schilder: Oval; margins well developed and prominent; base convex; teeth marked with brown; distinct dark blotch on inner lip; dorsum dark with white spots, actually the openings in a very dense reticulation; margins spotted. See *C. depressa*. Central Pacific Ocean. 33.3-89.1 (63.0) mm.

ZEBRA Linnaeus: Inflated, cylindrical; inner lip strongly projecting; anterior aperture relatively narrow; interior purplish; teeth marked with brown, relatively few; lateral spots ocellated. See *C. cervus, C. cervinetta*. Western Atlantic Ocean. 50.5-90.6 (83.2) mm.

NIGROPUNCTATA Gray: Inflated, elongate; tips prominent; dorsum heavily mottled with dark spots and blotches, becoming small black spots laterally; base whitish with small closely spaced black spots laterally. See *C. annettae.* Galapagos Islands and adjacent Ecuador and Peru. 23.3-36.1 (36.1) mm.

ACHATIDEA Sowerby: Oval; posterior tips not produced; teeth weak, numerous; base white, with pale orange pigment encroaching from the margins; margins orangish; dorsum heavily mottled with brown, often showing traces of transverse bands. See *C. langfordi.* Mediterranean Sea to West Africa. 23.5-40.0 (30.1) mm.

FRIENDII Gray: Highly variable in shape and color; tips extremely well developed, often continuous with margins of shell to produce small "wings"; usually moderately elongate-oval; aperture curved posteriorly; teeth absent over most of inner lip; base commonly black; dorsum cream with variable mottling, often nearly entirely dark. See *C. venusta, C. marginata*. Western and southern Australia. 60.5-92.3 (82.2) mm.

GUTTATA Gmelin: Teeth very long, extending onto the dorsal margins and tips; teeth stained with reddish; dorsum orangish with white spots of variable size. Eastern Indian Ocean to Melanesia. See *C. ocellata, C. lamarckii.* 62.8-69.0 (62.8) mm.

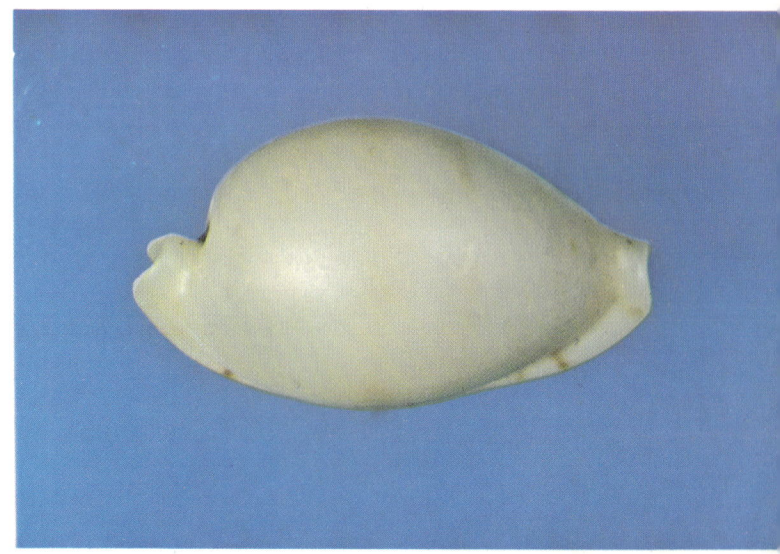

MARTINI Schepman: Elongated; tips prominent, especially posterior tips, which are very long and curved; teeth marked with brownish to purple; dorsum brownish with four narrow tan bands and with fine brown spots. See *C. rabaulensis, C. katsuae.* Ryukyus to central Pacific. 13.6-19.0 (17) mm.

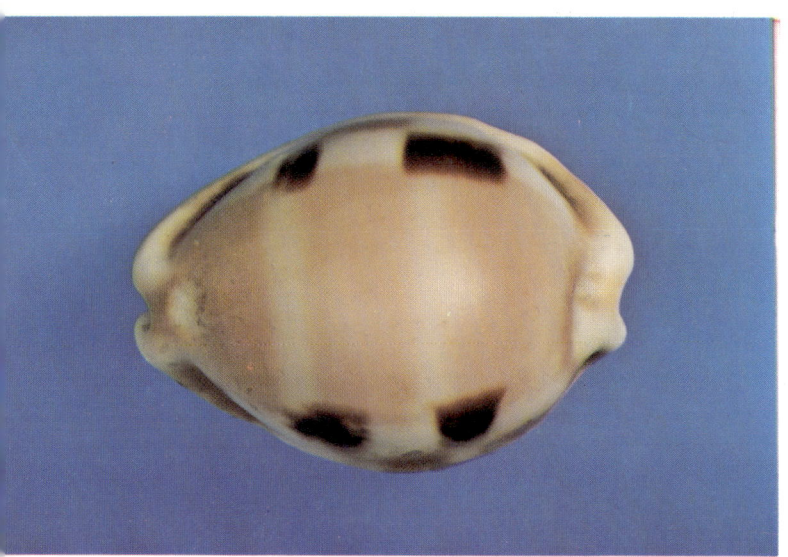

TESSELLATA Swainson: Oval; teeth heavy; broad tan bands alternating with broad white bands; sides with dark brown squares alternating with similar white squares; base marked with alternating tan and white. Hawaiian Islands. 15.2-50.1 (36.2) mm.

BRODERIPII Sowerby: Dorsum reticulated with reddish orange on a cream background, the size of the white spots variable. See *C. vitellus, C. nivosa.* South and central Indian Ocean. 66.4-92.8 (66.4) mm.

HADDNIGHTAE Trenberth: Dorsal spots large to small, relatively indistinct; lateral spotting absent or nearly so; elongate, tapering. See *C. cribraria*. Southwestern Australia. (34.0) mm.

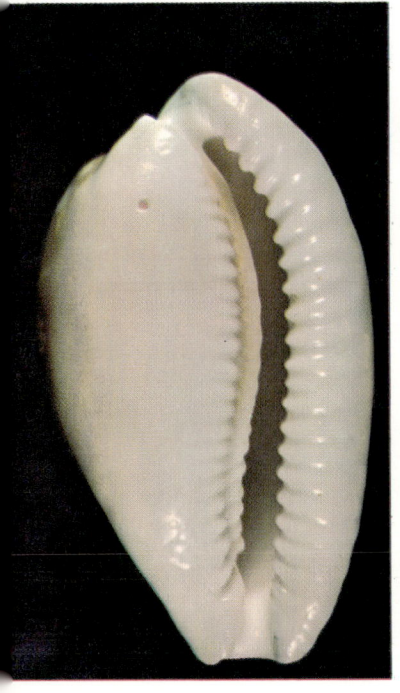 

ISABELLA Linnaeus: Cylindrical; tips of shell weakly produced, orange; grayish above with narrow white bands and variable number of small black streaks or spots usually arranged in lines; base pure white. See *isabellamexicana*. Indo-Pacific. 11.1-53.8 (33.8) mm.

MAUIENSIS Burgess: Small, rounded, inflated, with prominent twisted tips; dorsum smooth; cream with small brown spots especially at margins; spire blotch absent. See *C. margarita, C. cicercula.* Hawaiian Islands. 10-15 (13.0) mm.

EROSA Linnaeus: Shell oval; teeth heavy, those of outer lip long; two dark blotches on midside of shell, extending onto the base; basal spots few, light brown. See *C. nebrites*. Indo-Pacific. 20.7-56.3 (36.8) mm.

CRIBRARIA Linnaeus: Dorsal light spots usually large, sharply outlined; lateral spotting inconspicuous or absent; tapered shape. See *C. catholicorum, C. esontropia, C. haddnightae, C. cribellum.* Indo-Pacific. 18.4-36.7 (24.9) mm.

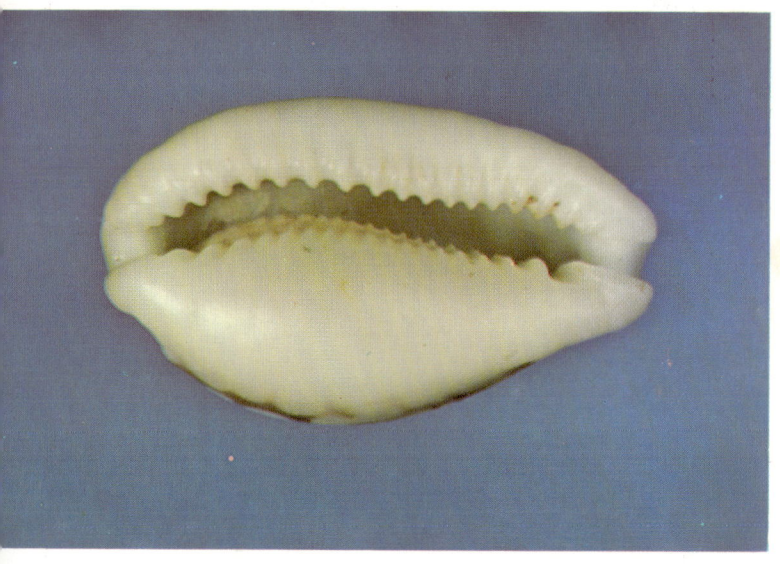

CRUICKSHANKI Kilburn: Greatly inflated, the outer lips slightly projecting; shell appears almost oval; both lips with teeth; outer lip with strong callus; creamy white, sometimes with brown spots; tips apricot. See *C. fuscorubra*. South Africa. 27.0-31.7 (27.0) mm.

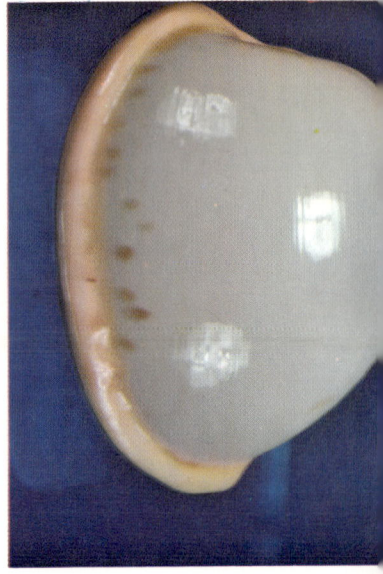

HUNGERFORDI Sowerby: Oval; anterior tips produced, posterior tips produced or not (inner weaker, outer stronger); base and sides whitish; dorsum lightly mottled with pale brown; lateral area heavily spotted with dark brown to black, the spots often joining to form a line separating the dorsum from the margins; traces of dorsal bands may be present. See *C. hirasei, C. midwayensis*. Japan and Queensland, Australia. 28.3-40.4 (34.9) mm.

ESONTROPIA Duclos: Dorsal light spots distinct, often crowded; lateral brown spots distinct, light, relatively large; tapers at both ends. See *C. cribellum, C. gaskoini, C. cribraria*. Mauritius only. 17.6-34.9 (23.2) mm.

NUCLEUS Linnaeus: A rough cowry heavily studded with nodules; relatively elongate; tips projecting; glossy yellow-brown. See *C. granulata*. Indo-Pacific. 14.9-28.7 (28.7) mm.

LANGFORDI Kuroda: Oval; tips produced; teeth weak, numerous; base and margins orange-tan; dorsum evenly and heavily mottled with darker brown over orange-cream background; sides not spotted. See *C. achatidea, C. hirasei*. Japan (right) and Queensland, Australia (left). 41.4-56.6 mm.

PULICARIA Reeve: Oval to elongated; margins of shell nearly parallel; tips and margins well developed; white, the dorsum with four brown bands interrupted to produce regular squarish spots; fossular area strongly developed. See *C. bicolor, C. reevei*. Southwestern Australia. 15.3-19.4 (18.7) mm.

LEUCODON Broderip: Heavy, greatly inflated; tips short but prominent; teeth very heavy, uniform in strength from anterior to posterior aperture; dorsum orangish brown with large but indistinct white spots; base tan to white. Central Indian Ocean to the western Pacific. 75.0-82.6 (82.6) mm.

LAMARCKII Gray: Elongate-oval; margins and tips prominent; anterior teeth of outer lip elongate, projecting to dorsum; tips with dark spots and streaks above; dorsum tan with scattered white spots and a few ocellated spots; base with brown spotting on the sides. See *C. labrolineata, C. miliaris, C. ocellata.* Indian Ocean. 28.0-49.4 (42.4) mm.

Two specimens of *C. friendii* subspecies *thersites*, variety *contraria*. *C.f. thersites* is found in South Australia and differs from typical *C.f. friendii* in being pale ventrally; variety *contraria* is pale dorsally as well in most specimens.

RASHLEIGHANA Melvill: Oval, inflated, left side with a strong, sharp callus; teeth fine, numerous; dorsum reddish or yellowish tan, variously mottled; sides with small brown spots which usually extend over about half of base. Very close to *C. teres.* Hawaiian Islands. 14.6-28.4 (23.1) mm.

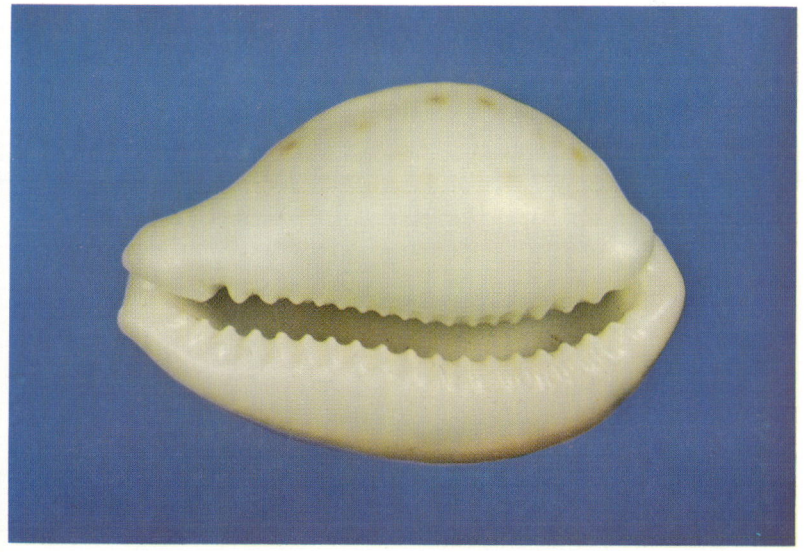

SEMIPLOTA Mighels: Dorsum smooth, not nodulose; anterior and posterior teeth of inner lip produced; reddish brown with numerous small white spots. See *C. staphylaea, C. limacina*. Hawaiian Islands. 7.2-26.0 (16.8) mm.

STAPHYLAEA Linnaeus: Dorsum with distinct nodules; shell globular in shape; teeth of inner lip extend across entire base; blackish to milky white, the nodules lighter; tips orangish. See *C. limacina, C. semiplota.* Indo-Pacific. 12.4-26.6 (19.3) mm.

CICERCULA Linnaeus: A small, round, inflated shell with prominent tips; dorsum granulate (or only sides granulate); a brown blotch over the spire and numerous small brown spots dorsally; base without dark blotches. See *C. bistrinotata, C. globulus, C. mauiensis.* Indo-Pacific. 10.3-20.5 (20.4) mm.

MIDWAYENSIS Azuma & Kurohara: Very similar to *C. hungerfordi*, but smaller, anterior teeth of inner lip fused, and outer teeth more numerous; pattern less distinct and with a more prominent central blotch. Southern Japan and China Sea. See *C. lisetae, C. hungerfordi*. 13.5-21.5 (19.0) mm.

SCURRA Gmelin: Inflated, cylindrical; teeth marked with brown; purplish brown dorsally with an open reticulate pattern; sides and margins with distinct dark spots. See *C. cervus, C. arabica*. Indo-Pacific. 27.0-52.6 (43.6) mm.

OCELLATA Linnaeus: Elongate-oval; margins and tips prominent; teeth heavy, stained with brown; base heavily spotted on sides; margins and tips with brown streaks; dorsum brown, with scattered white spots and ocellated dark spots of both large and small size. See *C. labrolineata, C. gangranosa, C. lamarckii*. Northern Indian Ocean. 19.4-30.2 (30.2) mm.

LUCHUANA Kuroda: Oval-elongate; teeth heavy, those on inner lip long; spire with a brown spot in adults; dorsum mottled with brown and with four broad brown bands irregular and indistinct against background; anterior tip pale, without brown smudges. See *C. pallidula, C. summersi, C. interrupta*. Ryukyu Islands. 16.6-23.2 (19.0) mm.

TERES Gmelin: Elongate-oval to elongate; tips moderately prominent; teeth numerous; dorsum brownish with three darker bands, the middle band usually darkest at the center and forming a dorsal blotch; sides with large distinct spots barely visible from ventral view. See *C. subteres, C. rashleighana*. Indo-Pacific. 13.6-39.6 (28.3) mm.

MARIAE Schilder & Schilder: A small, round, inflated shell; white with large circular brown spots over entire dorsum, the spots usually outlined with darker brown. Pacific Ocean. 9.4-20.0 (16.9) mm.

PORARIA Linnaeus: Broadly oval, the margins and tips not produced; dorsum brownish, heavily tinged with purplish, as is the base; numerous small white spots over dorsum, most outlined with a narrow brown ring; spotting on base and sides not prominent or dark. See *C. albuginosa, C. marginalis*. Indo-Pacific. 12.6-24.6 (17.0) mm.

Cypraea fuscorubra has proved to be a quite variable and confusing species. The shape, pattern (of beach shells), and number of teeth are very variable. The shell shown here was formerly called *gondwanalandensis,* but that name is a strict synonym of *fuscorubra,* as both names were proposed to replace *C. similis* (preoccupied).

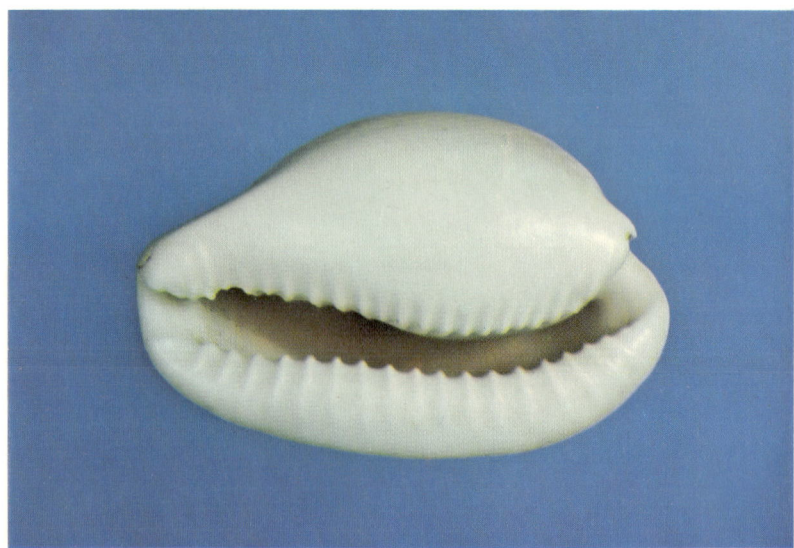

MARGINATA Gaskoin: Elongate-oval; tips well developed, continuous with strongly developed, frilly margins; aperture curved posteriorly; base usually cream to white with distinct spotting laterally; dorsum cream to white with small spots and mottlings or a large central dark dorsal figure. See *C. venusta, C. friendii*. Western and southern Australia. 51.6-62.4 (59.3) mm.

MILIARIS Gmelin: Elongate-oval; tips prominent; anterior teeth of outer lip elongate, projecting to dorsum; base, margins, and tips white, without spots or streaks; dorsum tan with numerous round white spots. See *C. lamarckii, C. eburnea, C. ocellata.* Western Pacific Ocean and northwestern Australia. 27.3-44.0 (33.4) mm.

BERNARDI Richard: Dorsal spots indistinct, small; lateral spotting absent or neary so; broadly tapering. See *C. kingae, C. cribraria.* French Polynesia. 12mm.

FELINA Gmelin: Variable; oval to elongate-oval; tips moderately prominent, marked with four large dark spots; sides with large dark spots; dorsum grayish to blue or greenish, usually with indications of two light bands under the heavy fine dark mottling. Indo-Pacific. 16.7-25.4 (20.6) mm.

ENGLERTI Summers and Burgess: Inflated, the margins indistinct and not dark brown; anterior large light spot absent; teeth not very heavy; base convex. See *C. caputdraconis*. Easter Island. 22.8-24.4 (23.5) mm.

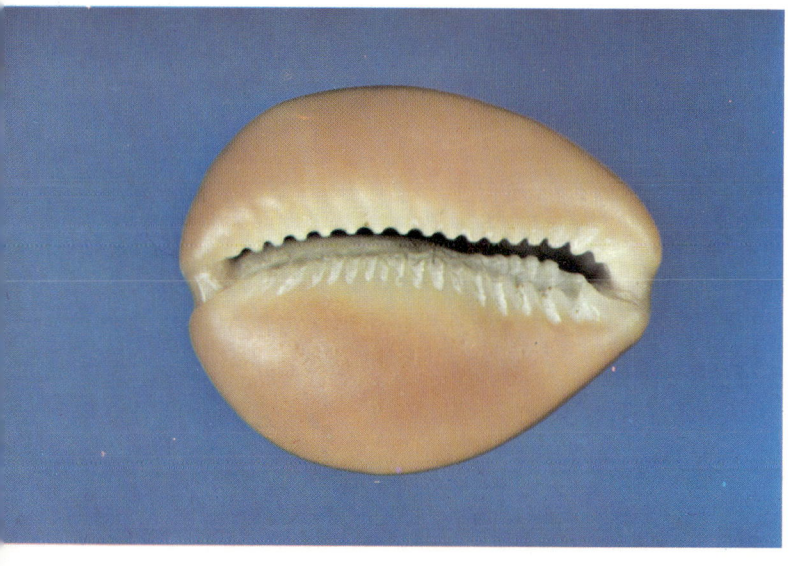

PORTERI Cate: Oval, inflated; margins heavy; tips moderately prominent; aperture nearly straight, with numerous but deeply cut short teeth; background of dorsum and base yellowish to apricot, dorsum overlaid by tan to violet spots and mottlings; base without spots. See *C. langfordi.* China Seas and Philippines. 45-58 (47.6) mm.

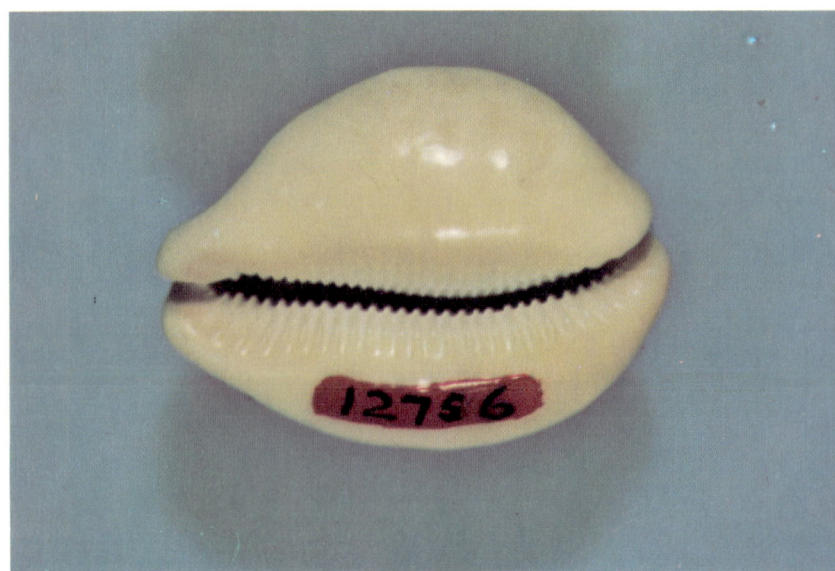

CATHOLICORUM Schilder & Schilder: Dorsal spots distinct, crowded; lateral spotting small, distinct; shape moderately tapering. See *C. cribraria, C. gaskoini*. New Britain, Solomons, New Hebrides. 14.0-21.3 (14.0) mm.

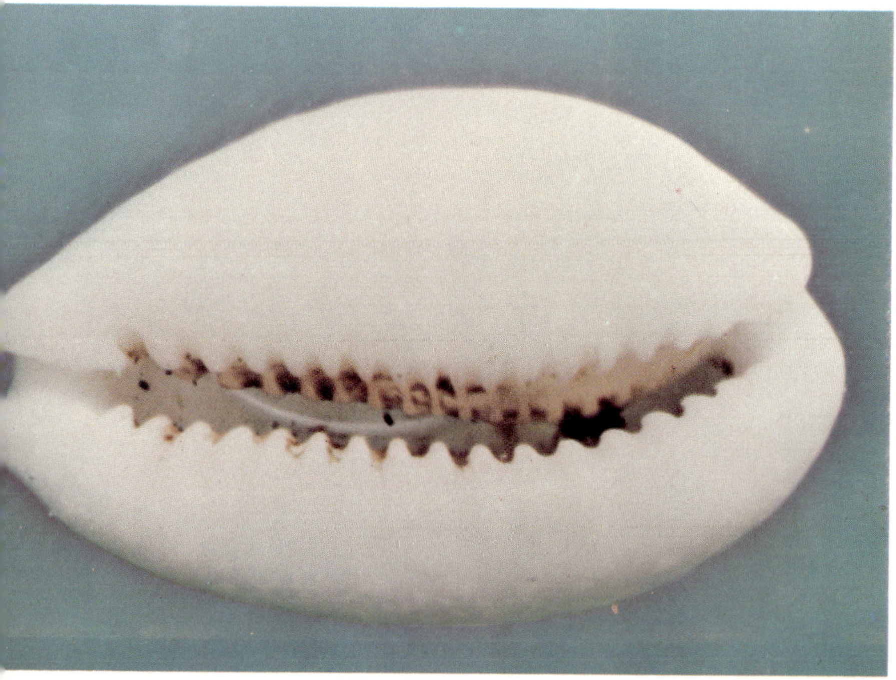

MARGINALIS Dillwyn: Oval, the margins and tips prominent and produced in adult shells; dorsum and base distinctly purplish-tan; sides with prominent brown spots extending onto base; dorsum with small white spots and ocellated spots. See *C. poraria, C. albuginosa, C. ocellata.* Western Indian Ocean. 21.5-31.6 (21.5) mm.

CAMELOPARDALIS Perry: Brown with distinct white spots irregularly scattered and not "swirled"; tips white; teeth of inner lip stained black. See *C. vitellus, C. nivosa*. Red Sea and Gulf of Aden. 31.0-78.0 (53.0) mm.

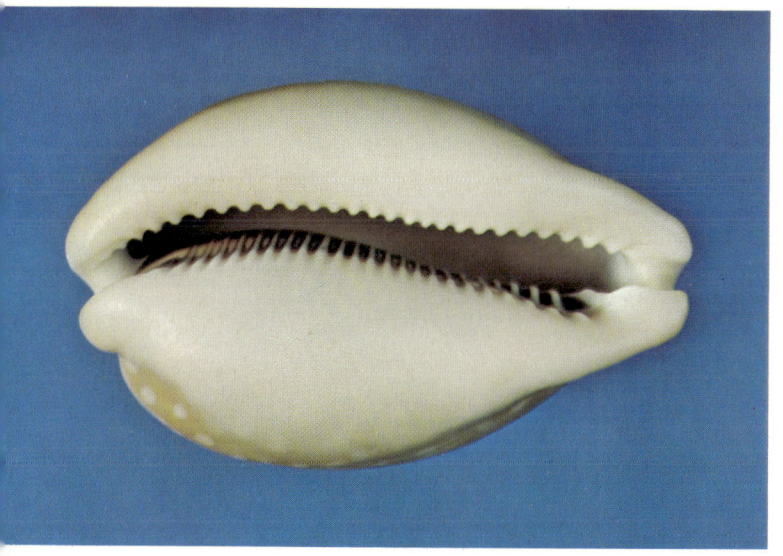

TERAMACHII Kuroda: Oval; tips strongly produced; aperture strongly curved posteriorly; teeth on inner lip weak, virtually absent near center; buffy tan, lightly spotted or mottled dorsally, with broad brown marginal bands overlying the spotted pattern. See *C. hirasei, C. armeniaca.* Japan and New Caledonia. 55-75 (69.4) mm.

TALPA Linnaeus: Cylindrical; base black; above golden yellow with four brown bands. See *C. exusta.* Indo-Pacific. 22.6-93.0 (52.8) mm.

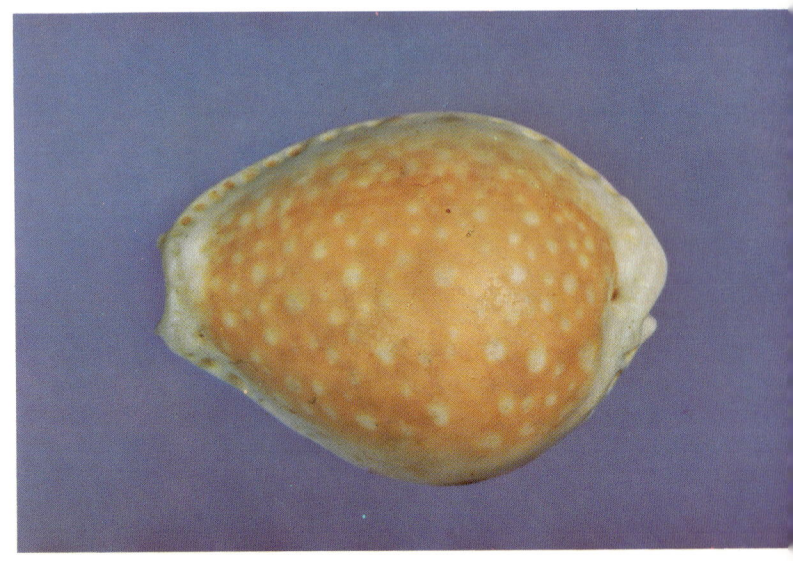

CERNICA Sowerby: Oval, inflated; anterior teeth of outer lip and posterior teeth of inner lip greatly prolonged; base white; margins with a row of irregular brown spots, sometimes indistinct; dorsum yellowish to orange or brown with scattered poorly defined white spots. See *C. spurca*. Indo-Pacific. 11.3-30.2 (23.0) mm.

Cypraea friendii jeaniana varies from nearly all dark brown to uniformly white. Notice the distinct columellar teeth. Exact range unknown, but presumably northwestern Australia.

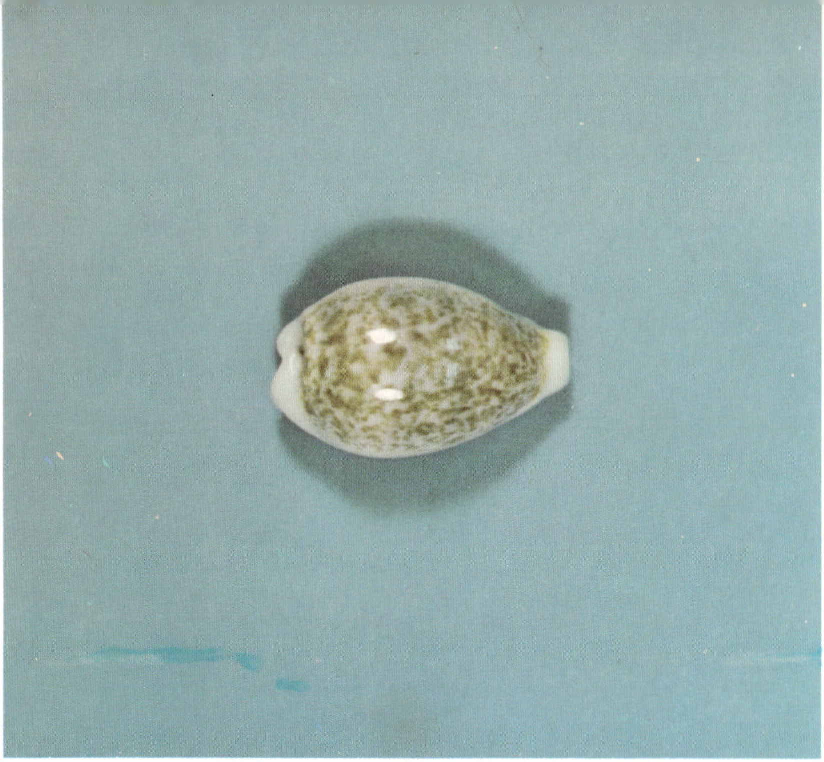

The holotype of *C. dayritiana,* **the** specimen on which the concept of this species must be based.

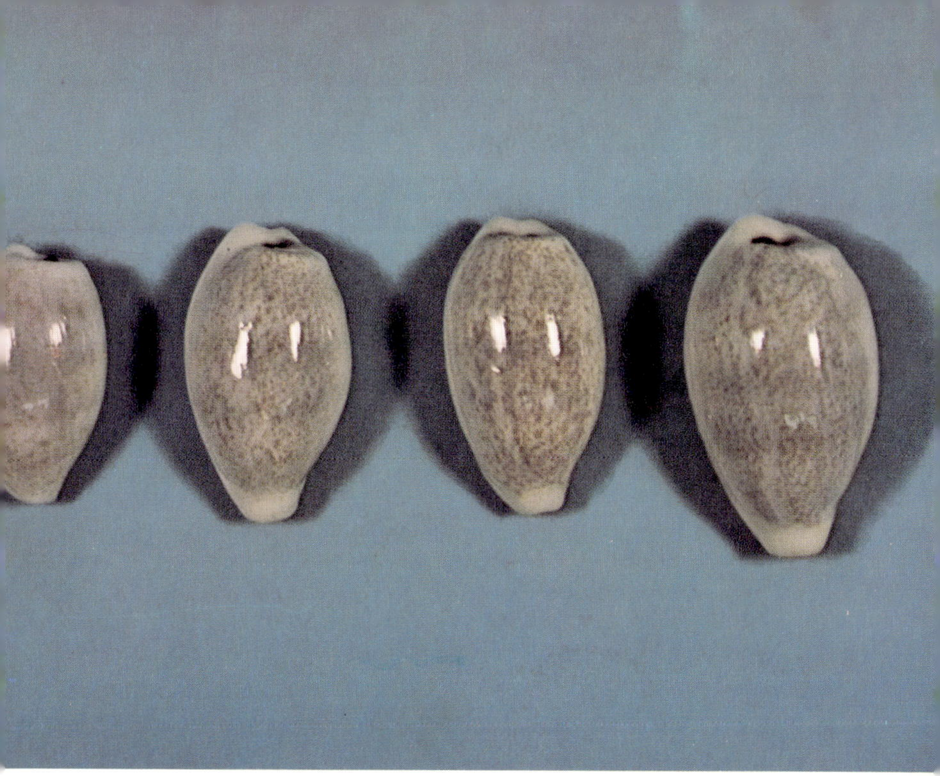

Above, a series of *C. luchuana* showing variation in shape and pattern; below, a ventral view of a small *C. luchuana*. All cowries are variable, many extremely so, in shape and color and often vary considerably with age.

CAPUTDRACONIS Melvill: Similar to *C. caputserpentis*, but less definitely margined; area between teeth brown; base convex; dorsal light spots smaller, more numerous. Easter Island. 17.0-45.0 (36.5) mm.

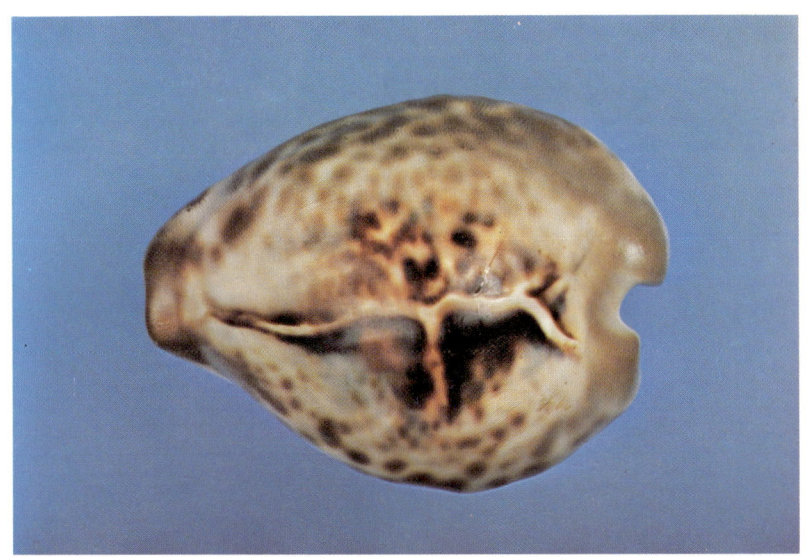

TEULEREI Cazenavette: Margins of adult shell depressed, irregular; teeth absent; edges of base with large brown spots; irregular dorsal figure. See *C. mus, C. fultoni*. Red Sea to Gulf of Oman. 45.9-51.0 (45.9) mm.

KATSUAE Kuroda: Elongate; posterior tips poorly produced, anterior tips narrow, prominent; teeth fine, stained with brown or not; base white; lateral spots present. (This is a decorticated specimen.) See *C. rabaulensis, C. martini, C. beckii.* Japan to the Philippines. 18-23 mm.

FIMBRIATA Gmelin: Cylindrical; tips of aperture purplish to slate; base not spotted; dorsum with a few small irregular spots; middorsal band darkest at edges, weak in center. See *C. gracilis, C. minoridens, C. serrulifera, C. hammondae.* Indo-Pacific. 8.0-21.0 (12.7) mm.

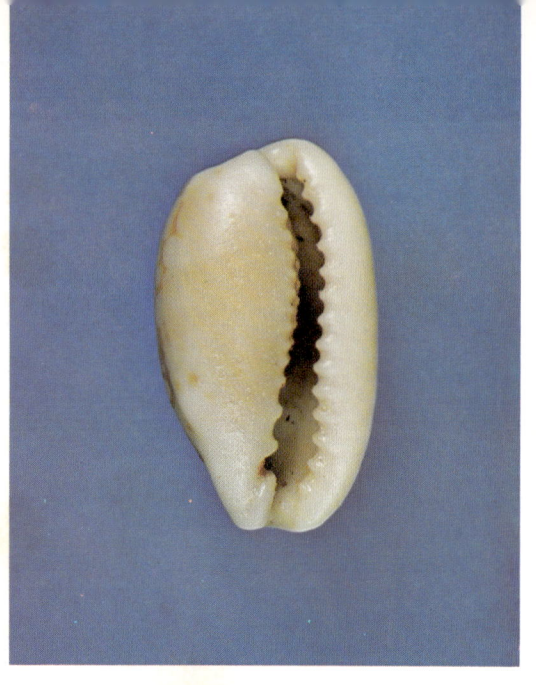

CRIBELLUM Gaskoin: Dorsal light spots distinct, crowded; lateral spotting absent or nearly so; shape cylindrical, not tapering at ends. See *C. esontropia*. Mauritius and Reunion only. 12.6-16.6 (13.3) mm.

GASKOINI Reeve: Dorsal spots distinct, often crowded; lateral spots small, numerous; moderately tapered to distinctly oval. See *C. cumingii*. Hawaiian Islands. 11.2-27.4 (13.6) mm.

CAPUTSERPENTIS Linnaeus: Moderately domed, the sides forming distinct margins; base flat or concave; teeth large, the areas between them whitish; chocolate brown with a large light spot anteriorly and posteriorly, middorsal area heavily light spotted. See *C. caputdraconis, C. englerti*. Indo-Pacific. 14.8-39.2 (32.5) mm.

Even rare shells are variable. Notice that this specimen of *C. rosselli* is much less angular marginally than the other specimen illustrated.

RABAULENSIS Schilder: Elongated; tips prominent, the anterior tips flared, wide and sharp; teeth not stained with brown; dorsum brown with four broken darker bands; base tan, with marginal spots. See *C. martini, C. katsuae.* Philippines to New Guinea. 20-25mm.

ANNETTAE Dall: Inflated, elongate; interior purplish; tips not prominent; dorsum heavily mottled with dark spots and blotches, only a few extending onto base; base tinged with brown, the teeth tipped with white. See *C. nigropunctata, C. cervinetta.* Gulf of California; Panama to Peru. 25.4-48.0 (36.0) mm.

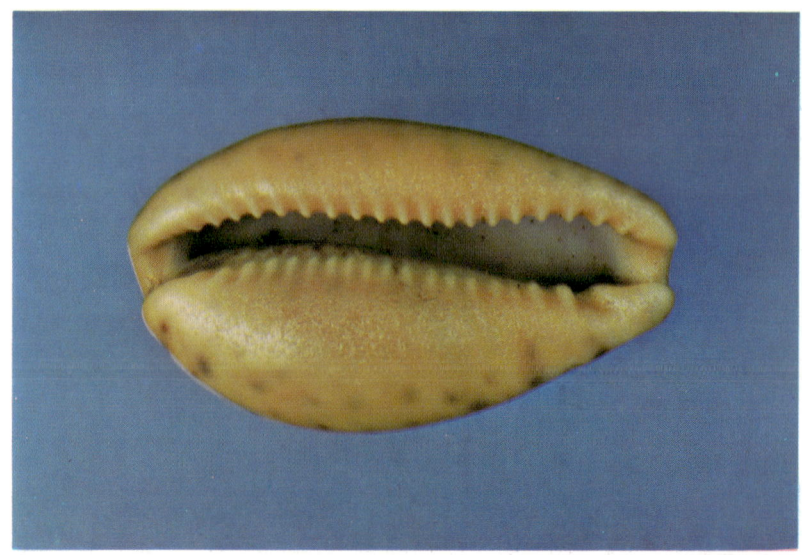

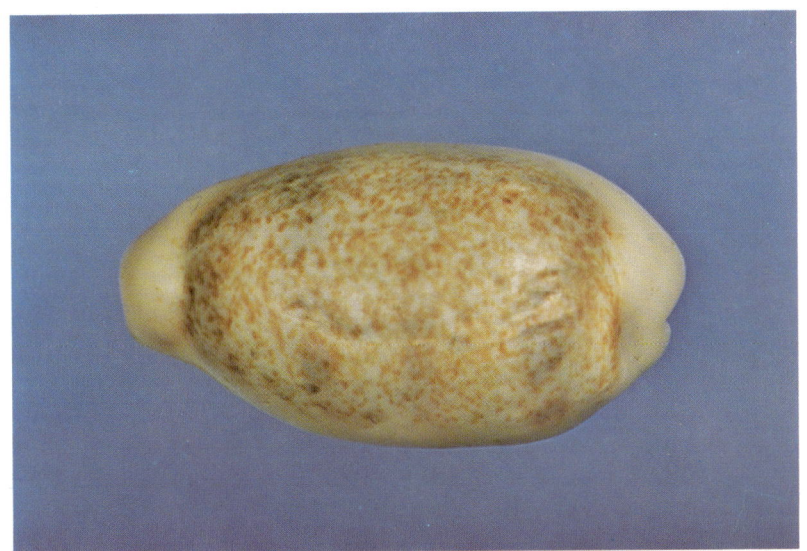

PALLIDULA Gaskoin: Relatively elongate; teeth heavy, numerous, those on inner lip long; dorsum with four equally spaced brown bands on mottled background, the bands usually continuous and strong, but occasionally weak; anterior tip smudged above with brown. See C. luchuana, C. interrupta, C. summersi. Western Pacific Ocean and northwestern Australia. 16.4-23.4 (21.9) mm.

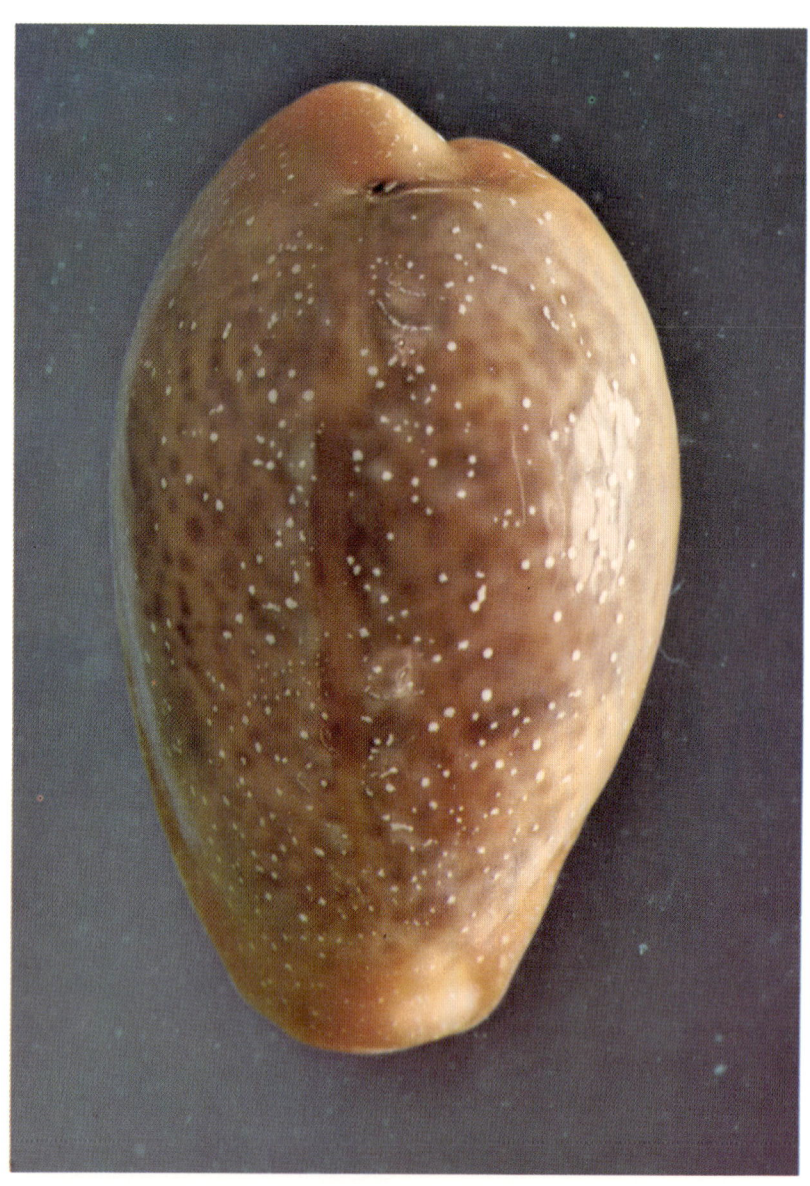

BREGERIANA Crosse: Similar to *C. walkeri* in basic shape and color pattern; base black to bright red; dorsal pattern often obscured by heavy, fine mottling; base, sides, and sometimes dorsum with small embedded white spots (most obvious in water). See *C. walkeri*. New Caledonia and Fiji to the Solomons. 16.5-30.0 (18.0) mm.

C. bregeriana, ventral view.

EGLANTINA Duclos: Cylindrical; not as inflated as related species; base convex; teeth marked with brown; small but distinct blotch on spire; dorsum with small reticulations not interrupting longitudinal lines as a rule. See *C. arabica, C. grayana, C. histrio.* Pacific Ocean, northwestern Australia. 38.1-69.2 (48.8) mm.

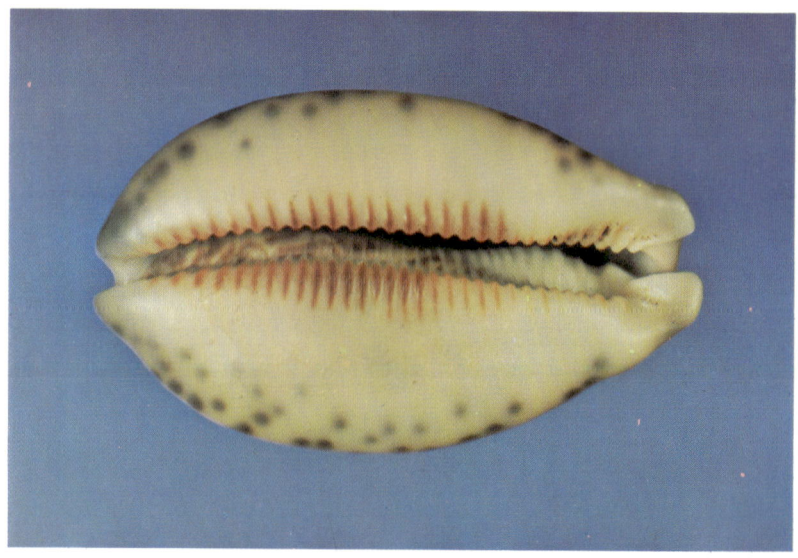

HELVOLA Linnaeus: Broadly oval; teeth heavy, long; base usually orange to brown; dorsum with numerous small scattered white spots and a few larger dark brown spots; sides usually orange brown, the tips violet. See *C. citrina, C. thomasi.* Indo-Pacific. 9.6-36.2 (19.6) mm.

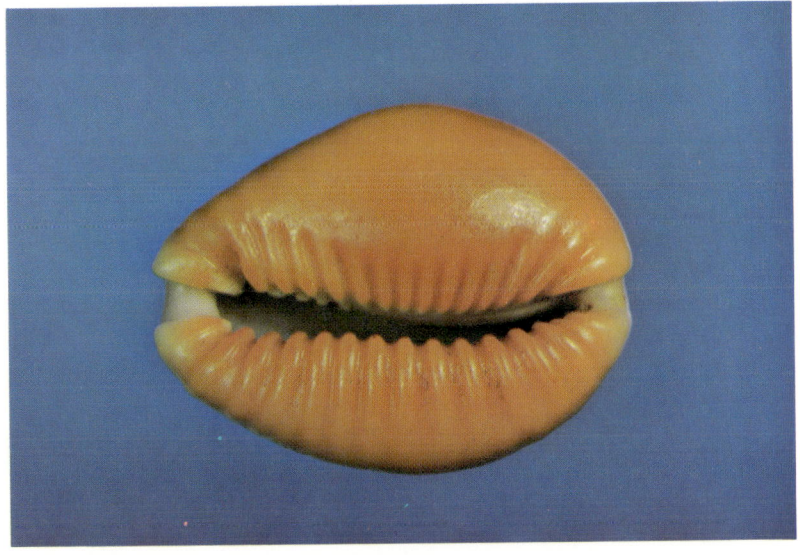

This is the holotype of *Cypraea steineri*, now believed to be an inflated freak of *C. coxeni*. It agrees in pattern and teeth with that species but the shape is obviously distinct. Freaks such as this occur in many species and easily lead to confusion.

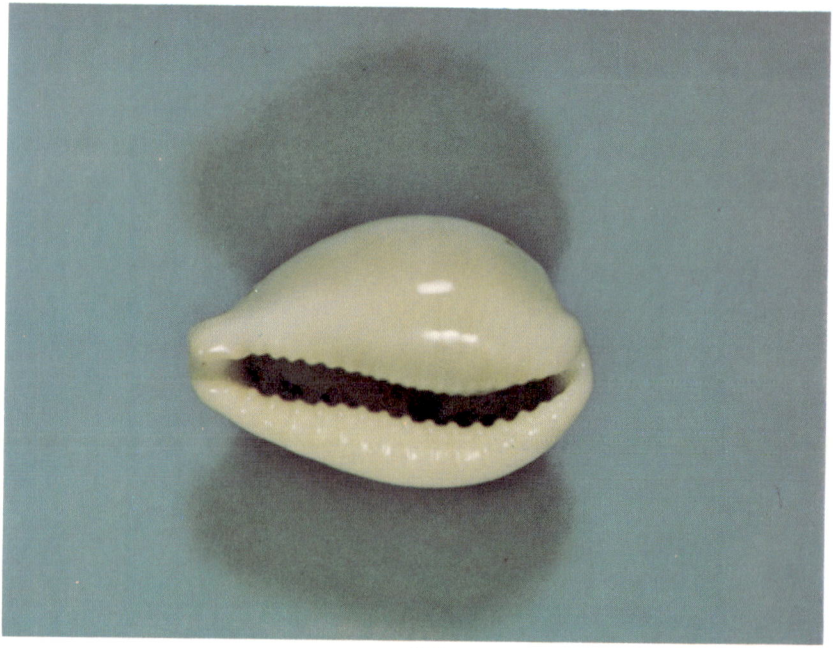

LISETAE Kilburn: Extremely similar to *C. midwayensis* in size, shape, and pattern, but with finer teeth and somewhat more inflated shape. Southern Mozambique and Natal. 14-17mm.

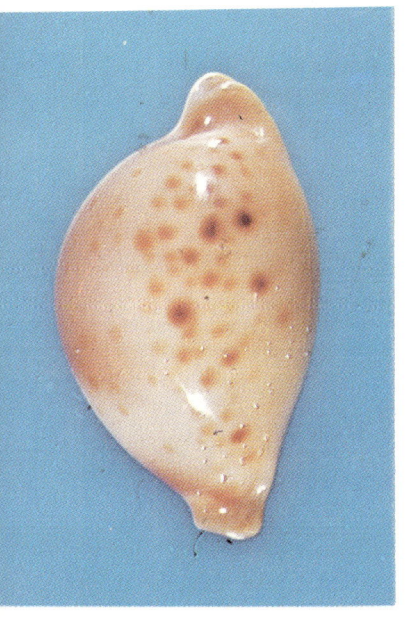

VITELLUS Linnaeus: Brown with distinct white spots arranged in a "swirled" pattern; sides with fine vertical striations; aperture nearly straight. See *C. camelopardalis, C. nivosa.* Indo-Pacific. 26.1-79.5 (49.6) mm.

CITRINA Gray: Broadly oval; base orange with a long brownish stain on inner lip; margins and tips orange; dorsum brownish with numerous small white spots, no dark spots. See *C. helvola*. South Africa. 16.7-30.0 (24.0) mm.

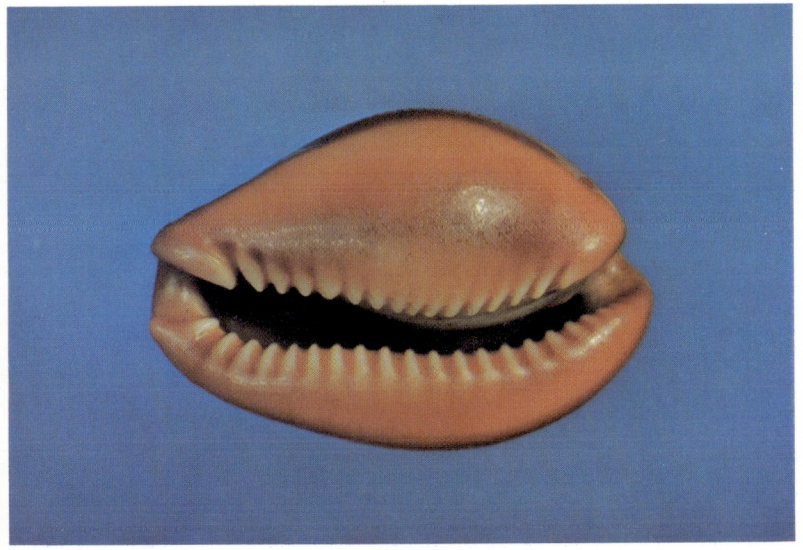

SAKURAI Habe: Elongate-oval; tips produced; aperture slightly curved posteriorly; teeth strong, numerous; tan dorsally and laterally with a large dark brown mottled figure dorsally and high on sides; no spots on sides; base white. See *C. hirasei, C. teramachii*. South China Sea. 44 mm.

 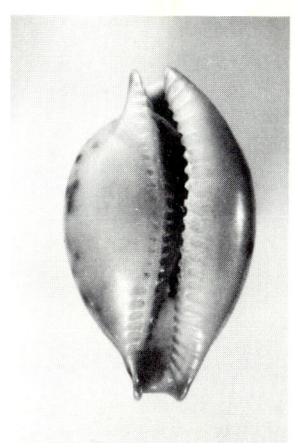

MARICOLA Cate: Unique holotype from the Philippines. Extremely close to *margarita* but with fused columellar teeth. Perhaps aberrant. 14mm.

FERNANDOI Cate: Holotype from the Philippines. Very similar to *xanthodon* and *vredenburgi* but more slender and with longer and narrower terminals. Perhaps a mislocalized shell or distinct Philippine race of *vredenburgi*. 19mm.

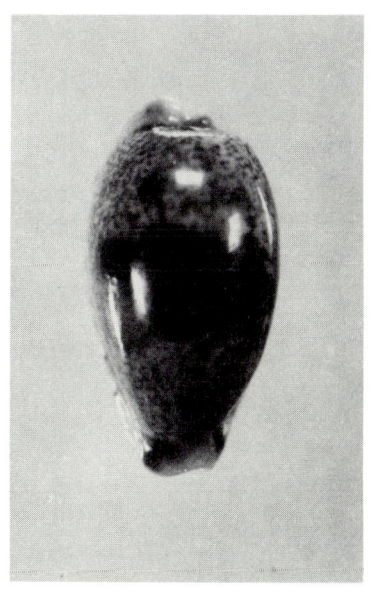 

SYNOPSIS OF
COWRY SUBSPECIES AND VARIETIES

The problem of subspecies and other variants in cowries is a difficult and co..-plicated one that has not been directly approached by recent authors. Although cowries are obviously very variable and have been divided into subspecies with discrete geographical ranges, there is also much individual variation and variation from population to population. It is sometimes possible to find specimens of many "subspecies" of a species on a single reef, as for example in Philippine *Cypraea ziczac, C. caputserpentis,* and *C. helvola.* Since dealers and collectors will persist in using "subspecies names" for varieties and minor races as well as real subspecies regardless of what anyone says, we have included this section as a guide to the myriad of names.

We have attempted to present a classification of cowry infraspecific variants (variation below the species level) based on specimens examined, the literature, and personal opinion. The species are presented in alphabetical order followed by the author's name, date of description, minimum and maximum lengths of adults,* and a page reference to a photo in the main body of the book. *Subspecies* (readily distinguishable variants that occupy *definite geographic ranges*) and major *varieties* (distinguishable variants that are *not localized* but are found in scattered localities or only *tend* to be distinguishable in certain geographic areas) are listed and defined. Many lesser variants (*races*) are sold that may have characters appearing distinctive at first glance but which are either ecological in nature, exceedingly variable, or typical of very restricted populations. These are not considered as meriting their own names and are listed under *Familiar synonyms* along with their general distribution as given by Schilder. Most cowry specialists collect geographically anyway and for some reason prefer to call a shell "*walkeri continens*" rather than "East Australian *walkeri*" regardless of the true validity of the name *continens.*

It must be remembered that even subspecies are not always constant and may vary somewhat from population to population or even from individual to individual. The characters given are those that seem most constant and least subject to variation over the entire range. Slender-inflated, dark-light, giant-dwarf, callused-nacred, and unusually colored specimens occur in almost any large sample and are usually not given distinctive names but considered to be within the normal variation of a subspecies just as sex, height, weight, and genetic deformities are considered part of normal human variation. This list is meant only as a guide, not a final solution.

*These lengths are taken, with modification, mostly from an article by R. H. Jones and H. G. Lee appearing in *Hawaiian Shell News,* 25 (11), 1978.

ACHATIDEA Sowerby, 1837 [22-42mm] **191**
No variants recognized.
Familiar synonyms: *inopinata, longinqua, oranica.*
ALBUGINOSA Gray, 1825 [12-34mm] **134**
No variants recognized.
Familiar synonym: *nariaeformis*—Galapagos to Ecuador.
ALGOENSIS Gray, 1825 [16-28mm] **156**
No variants recognized. Hybrids with *edentula* have been called *cohenae.*
ANGUSTATA Gmelin, 1791 [19-36mm] **141**
Very variable in color and pattern according to ecological conditions, becoming paler with depth.
Familiar synonyms: *albata; emblema; moelleri; subcarnea; verconis.*

ANNETTAE Dall, 1909 [22-52mm] **246**
Two strong subspecies:
annettae annettae: Relatively narrow and cylindrical; posterior aperture weakly curved; anterior terminals usually dark above; Gulf of California.
annettae aequinoctialis: Broader and more oval; posterior aperture strongly curved; often a white spot on anterior terminals; Panama Bay to northern Peru.

ANNULUS Linnaeus, 1758 [10-34mm] **136**
Very variable in size and shape, but seemingly only reflecting ecological conditions. Two subspecies can be recognized:
annulus annulus: Margins of shell not inflated; Indo-Pacific.
Familiar synonyms: *camelorum*—W. Indian O.; *harmandiana*— Malaysia; *noumeensis*—W. & C. Pacific O.; *scutellum*—N. & C. Indian O.
annulus obvelata (**132**): Margins of shell greatly inflated; French Polynesia.
Familiar synonym: *perrieri*—no gold ring.

ARABICA Linnaeus, 1758 [17-105mm] **184**
Two weak and intergrading subspecies:
arabica arabica: Rather cylindrical, the anterior terminals shorter and broader; base flat or concave; spire blotch absent; Indo-Pacific.
Familiar synonyms: *asiatica*—Japan to New Guinea; *brunnescens*—N.W. Australia; *dilacerata*—E. Indian O.; *gibba*—Malaysia; *immanis*—W. & C. Indian O.
arabica grayana (**180**): Distinctly humped posteriorly, the anterior terminals longer and more pointed; base more convex; spire blotch absent or weakly indicated; Red Sea and Arabia.

ARABICULA Lamarck, 1810 [16-35mm] **166**
Large and small variants occur.

ARGUS Linnaeus, 1758 [47-107mm] **113**
Blotches may be present or absent on the outer lip, the pattern of ocelli dense or sparse, and the shape cylindrical to inflated, but no variants are recognized.
Familiar synonyms: *concatenata; contrastriata*—W. Indian O.; *ventricosa*— Oceania.

ARMENIACA Verco, 1912 [54-121mm] **108**
Two strong subspecies:
armeniaca armeniaca: Anterior terminals shorter, placing the greatest height near midbody; anterior teeth relatively short and coarse; base tinted with orange; W. Great Australian Bight.
armeniaca hesitata: Anterior terminals distinctly longer, placing the greatest height posterior to midbody; anterior teeth finer and longer; base white, sometimes with tan blotch; S. E. Australia. Three varieties:
a. h. var. *typical:* Distinctly mottled dorsally; over 75mm long; entire range.
a. h. var. *howelli:* Dorsal pattern greatly reduced or absent; Victoria. (Caution: lack of pattern can be produced by baking.)
a. h. var. *beddomei:* Dorsal pattern mottled; small, under 80mm long; central New South Wales.

ARTUFFELI Jousseaume, 1876 [11-23mm] **119**
No variants recognized.

ASELLUS Linnaeus, 1758 [10-31mm] **102**
No variants recognized.
Familiar synonyms: *bitaeniata*—Melanesia; *latefasciata*—S.E. Australia; *vespacea*—Andamans to Japan & New Guinea.

AURANTIUM Gmelin, 1791 [58-117mm] **117**
No variants recognized.
Familiar synonym: *aurora.*

BARCLAYI Reeve, 1857 [22-28] **74**
No variants recognized.
BECKII Gaskoin, 1836 [7-16mm] **67**
No variants recognized.
BERNARDI Richard, 1974 [11-19mm] **225**
No variants recognized.

BICOLOR Gaskoin, 1849 [16-31mm] **79**
(= *piperata* of first edition; see Schilder, 1974, *Veliger*, 7(1): 37.) Variable in shape, solidity, and density of color, but no variants recognized.
Familiar synonyms: *euclia; occidentalis; reticulifera; wilkinsi.*
BISTRINOTATA Schilder & Schilder, 1937 [10-33m] **76**
Basal blotches may be very weak or absent. No variants recognized.
Familiar synonyms: *keelingensis*—E. Indian O.; *mediocris*—W. Pacific O.; *sublaevis*—C. Pacific O.
BOIVINI Kiener, 1843 [14-37mm] **130**
No variants recognized.
Familiar synonym: *amoena.*
BREGERIANA Crosse, 1868 [15-33m] **248**
Two pattern variants available: zonate dorsum and mottled dorsum.
BRODERIPII Sowerby, 1832 [66-104mm] **196**
No variants recognized.
CAMELOPARDALIS Perry, 1811 [31-81mm] **231**
No variants recognized.
CAPENSIS Gray, 1828 [24-38mm] **178**
Coloration and degree of development of the ridges varies, perhaps with extent of beach erosion. Hybrids with *edentula* have been called *amphithales* (**146**).
CAPUTDRACONIS Melvill, 1888 [17-45mm] **238**
Very close to *caputserpentis* and perhaps best considered a southern subspecies. No variants recognized.
CAPUTSERPENTIS Linnaeus 1758 [15-43mm] **243**
Highly variable in shape, details of pattern, color of base, and strength of the teeth. Warm-water populations are generally ovoidal, heavy, with pale bases and clearly defined terminal blotches. In cooler waters the shell becomes more cylindrical, lighter in weight, with the base tan and the terminal blotches less definite. In South Africa, S.W. Australia, and S.E. Australia these cool-water shells may be especially well defined and very distinct from warm-water shells. Since the variation appears to be clinal in nature (continuous over a long geographical range), no variants are recognized.
Familiar synonyms: *argentata*—Oceania, also used for pale-backed freaks; *caputanguis*—S.E. Australia; *caputophidii*—Hawaii; *kenyonae*—S.W. Australia; *mikado*—Japan; *reticulum*—Andamans to Japan and New Guinea.
CARNEOLA Linnaeus, 1758 [17-130mm] **142**
Highly variable in size and shape, development of callus, and presence or absence of a blue marginal line. Very large specimens (over 50mm) with nodules developed in the callus and with deformed radular teeth have been recognized as a full species, *leviathan* (**121**), but we feel that this name represents ecological or age variants rather than a valid species; the name may be retained as a variety if desired. In many or most localities large and small specimens co-exist with each other.
Familiar synonyms: *crassa*—Red Sea & Arabia; *gedlingae*—N.W. Aust.; *leviathan*—Oceania; *pretiosa, titan*—E. Africa; *propinqua*—Oceania; *sowerbyi*—Indian O.

CATHOLICORUM Schilder & Schilder, 1938 [10-23mm] **229**
Somewhat variable in spotting and over-all shape, but no variants recognized. The relationship of this species with *cribraria* and *cumingii* is uncertain.
CAURICA Linnaeus, 1758 [18-70mm] **99**
Highly variable in shape, pattern, development of marginal spots, over-all color, and strength of callus, but no variants recognized.
Familiar synonyms: *blaesa*—N.W. Australia; *corrosa*—N. Indian O.; *dracaena*—W. & C. Indian O.; *elongata*—E. Africa; *longior*—N. Australia; *nigrocincta; obscura; quinquefasciata*—Red Sea & Arabia; *thema*—Oceania.
CERNICA Sowerby, 1870 [10-37mm] **234**
Tends to occur in strongly callused small forms and larger less callused ones. No variants recognized.
Familiar synonyms: *marielae*—Hawaii; *maturata, ogasawarensis*—Ryukyus; *tomlini*—E. Australia & Melanesia; *viridicolor*—W. Australia.
CERVINETTA Kiener, 1843 [32-115mm] **168**
Large and small forms are about equally common.
CERVUS Linnaeus, 1771 [42-191mm] **185**
Large and small forms are about equally common.
CHILDRENI Gray, 1825 [12-30mm] **129**
No variants recognized.
CHINENSIS Gmelin, 1791 [7-52mm] **186**
Two subspecies recognized:
chinensis chinensis: Shape cylindrical to oval; base white to pale orange; usually distinct round colorless spots dorsally and dark spots at the margins; Indo-Pacific.
Familiar synonyms: *amiges*—Hawaii; *sydneyensis*—S.E. Australia; *tortirostris*—S. Africa; *variolaria*—C. Indian O.; *violacea*—E. Africa; *whitworthi*—N.W. Australia.
chinensis coloba **(137):** Broadly ovate with very heavy callus; base pale to dark orange; colorless round spots not developed dorsally and marginal dark spots weak or absent; N.E. Indian Ocean.
Familiar synonym: *greegori.*
CICERCULA Linnaeus, 1758 [8-23mm] **214**
No variants recognized.
Familiar synonym: *lienardi*—N. & C. Indian O.
CINEREA Gmelin, 1791 [15-42mm] **101**
No variants recognized.
CITRINA Gray, 1825 [13-30mm] **255**
Narrow and inflated forms occur.
CLANDESTINA Linnaeus, 1767 [8-26mm] **75**
Adult size (larger in Indian O.), density of dorsal zones (strongest at north and south), and development of dorsal zigzag lines (most prominent in Pacific O.) all vary. No variants recognized.
Familiar synonyms: *candida*—W. oceania; *extrema*—S.E. Australia; *moniliaris*—Andamans to Japan and New Guinea; *passerina*—S.E. Africa; *whitleyi*—N.E. Australia.
CONTAMINATA Sowerby, 1832 [8-16mm] **93**
No variants recognized.
Familiar synonyms: *distans*—W. Indian O.; *malaysiae*—Malaysia.
COXENI Cox, 1872 [14-29mm] **153**
Narrower shells with the mottling consolidated into large spots dorsally are often recognized as the variety *hesperina.* Inflated freaks have been described as *steineri* **(252).**

CRIBELLUM Gaskoin, 1849 [11-19mm] **242**
No variants recognized. The relationship of this species with *cribraria* is very close.
CRIBRARIA Linnaeus, 1758 [10-42mm] **201**
Variable in size, shape, size and arrangement of dorsal spots, and intensity of color. Albinistic individuals from Queensland are commonly called variety *melwardi*. No other variants recognized.
Familiar synonyms: *comma*—S.W. Indian O.; *exmouthensis, fallax*—W. Australia; *northi*—Fiji; *orientalis*—W. Melanesia; *zadela*—E. Australia.
CRUICKSHANKI Kilburn, 1972 [20-35mm] **202**
No variants recognized.
CUMINGII Sowerby, 1832 [9-30mm] **143**
Three widely intergrading variants:
cumingii var. typical: Smaller (under 20mm); teeth relatively fine and numerous; lateral spots fewer.
cumingii var. *astaryi:* Smaller (under 20mm); teeth coarser and fewer; lateral spots numerous; oval.
cumingii var. *cleopatra:* Larger (20-30mm); teeth coarser and fewer than typical; lateral spots numerous; elongated.
CYLINDRICA Born, 1778 [18-47mm] **152**
Two weak subspecies;
cylindrica cylindrica: Relatively narrow and cylindrical, the base flatter; columellar teeth somewhat longer; Pacific O.
Familiar synonym: *lenella*—E. Australia.
cylindrica sowerbyana: Somewhat broader and less cylindrical, the base a bit more convex; columellar teeth shorter; N.W. Australia.
Familiar synonym: *sista*.
DAYRITIANA Cate, 1963 [14-22mm] **118**
No variants recognized. Uncomfortably close to *pallidula*.
DECIPIENS Smith, 1880 [46-70mm] **165**
Two well defined variants:
decipiens var. typical: Base and margins uniformly dark brown; marginal spots not visible; shallow water.
decipiens var. *perlae:* Base and margins pale; marginal spots strong; deep water.
Additionally, a "contraria" phase that is nearly white with orange terminals occurs (= *joycae* of first edition, in error).
DECLIVIS Sowerby, 1870 [15-32mm] **145**
No variants recognized. Very close to *angustata*.
DEPRESSA Gray, 1824 [23-56mm] **181**
No variants recognized.
Familiar synonyms: *dispersa*—Indian O.; *gillei*—Polynesia.
DILLWYNI Schilder, 1922 [10-16mm] **84**
No variants recognized.
DILUCULUM Reeve, 1845 [11-36mm] **183**
Two minor variants:
diluculum var. typical: Larger; dorsal zigzag lines coarser and irregular; posterior terminal blotches present.
diluculum var. *virginalis:* Smaller, dorsal zigzag lines fine and complete; posterior terminal blotches absent.
EDENTULA Gray, 1825 [17-34mm] **148**
Considerably variable in development of dorsal spots, thickness of shell, and strength of teeth on outer lip. The name *alfredensis* has been used for heavier shells with a distinct middorsal blotch but seems to be meaningless.

Hybrids with *algoensis* (called *cohenae*) and *capensis* (called *amphithales*) have been described.

EGLANTINA Duclos, 1833 [35-80mm] **250**
No variants recognized.
Familiar synonyms: *couturieri*—Japan to New Britain; *perconfusa*—N.W. Australia.

ENGLERTI Summers & Burgess, 1965 [20-25mm] **227**
Large and small forms occur equally commonly.

EROSA Linnaeus, 1758 [16-71mm] **200**
Very variable in shape, callus development, and pattern, but no variants recognized.
Familiar synonyms: *chlorizans*—Melanesia; *lactescens*—E. Polynesia; *phagedaina*—Andamans to Japan & New Guinea; *purissima*—E. & N. Australia; *similis*—E. Africa.

ERRONES Linnaeus, 1758 [13-43mm] **104**
Variable in shape, color, and tooth development, but no variants recognized.
Familiar synonyms: *azurea*—N.W. Australia; *bimaculata*—N. Indian O.; *coerulescens*—S. Oceania; *coxi*—W. Melanesia; *magerrones; nimiserrans; proba*.

ERYTHRAEENSIS Sowerby, 1837 [13-27mm] **72**
No variants recognized. Probably a subspecies of *stolida*.

ESONTROPIA Duclos, 1833 [12-36mm] **204**
No variants recognized. Very close to *cribraria*.

EXUSTA Sowerby, 1832 [52-91mm] **107**
Color varies from pale to deep brown dorsally. Very close to *talpa*, probably just a subspecies.

FELINA Gmelin, 1791 [10-27mm] **226**
Very variable in size, shape, and pattern, but two weak subspecies can be recognized.
felina felina: Columellar teeth relatively longer; base creamy yellow to pale tan; E. Africa to Arabia.
Familiar synonym: *fabula*—Arabia.
felina listeri: Columellar teeth shorter, weaker; base white; E. & C. Indian O. and Pacific O.
Familiar synonyms: *melvilli*—Oceania; *pauciguttata*—Andamans to Japan and New Guinea; *velesia*—E. Australia.

FERNANDOI Cate, 1969 [19mm] **257**
Doubtful species. Very close to *vredenburgi*.

FIMBRIATA Gmelin, 1791 [7-21mm] **241**
There is much local and clinal variation, but the eastern and western extremes can be recognized as weak variants:
fimbriata var. typical: About 12mm long; usually pale; dorsal banding commonly indistinct; Indo-Pacific.
Familiar synonym: *marmorata*.
fimbriata var. *durbanensis:* Often over 14mm long; pale; dorsal band commonly with distinct zigzags; South Africa.
fimbriata var. *unifasciata:* Often only 10mm long; usually darker; dorsal band often sharp and nearly continuous; Polynesia.

FRIENDII Gray, 1831 [42-104mm] **192**
Three subspecies:
friendii friendii: Columellar teeth absent or nearly so; relatively narrow; base usually totally dark brown; S.W. Australia.
Two varieties:
f.f. var. typical: Shape narrow.

f.f. var. *vercoi:* Shape very broad, ovate.
friendii thersites **(210):** Columellar teeth absent or nearly so; distinctly broad and oval; base usually with pale areas along aperture; S.E. Australia.
friendii jeaniana **(235):** Columellar teeth well developed; broad and oval; color very variable, base varying from dark to light; N.W. Australia.
Additionally, deep-water shells of any subspecies tend to be more inflated, whitish to pale golden brown, and with orange terminals. Such shells are usually called variety *contraria*.

FULTONI Sowerby, 1903   [50-66mm]   **103**
No variants recognized.

FUSCODENTATA Gray, 1825   [24-43mm]   **170**
Varies somewhat in shape, with short and broad specimens uncommon. Familiar synonym: *coronata*.

FUSCORUBRA Shaw, 1909   [22-46mm]   **100**
Two variants may be recognized, but their status is uncertain:
fuscorubra var. typical: Shape rather oblong; aperture wider; fossular region concave in profile to some extent; shallower water.
fuscorubra var. *iutsui:* Shape inflated, globose; aperture narrower; fossular region straighter in profile; deeper water.
Familiar synonyms: *castanea; gondwanalandensis; similis.*

GANGRANOSA Dillwyn, 1817   [9-27mm]   **159**
Two weak variants:
gangranosa var. typical: Whitish nacre usually absent or poorly developed on the sides; terminal blotches usually larger.
gangranosa var. *reentsii:* Whitish nacre strongly developed on the sides; terminal blotches usually smaller.

GASKOINI Reeve, 1846   [10-27mm]   **242**
Tends to become broader and thicker with larger size. No variants recognized.

GLOBULUS Linnaeus, 1758   [9-24mm]   **86**
No variants recognized.
Familiar synonyms: *brevirostris*—W. & C. Indian O.; *sphaeridium*—Oceania.

GOODALLI Sowerby, 1832   [8-20mm]   **71**
Two variants:
goodalli var. typical: Broader; terminal spots weak or absent, pale brown.
goodalli var. *fuscomaculata:* More cylindrical; terminal spots large, blackish.

GRACILIS Gaskoin, 1849   [9-30mm]   **123**
Although there is great variation in adult size, color, size and strength of dorsal spot, and color of terminals, no variants are recognized.
Familiar synonyms: *hilda, irescens*—N.W. Australia; *japonica*—N. China Sea; *macula*—E. Australia; *notata*—Red Sea & Arabia.

GRANULATA Pease, 1862   [15-43mm]   **140**
Two subspecies:
granulata granulata: Dorsal groove shallow and partially interrupted by tubercles; rosy to grayish above and below; Hawaiian Islands.
granulata cassiaui: Dorsal groove deep and with continuous sides; purplish above and orange below; Marquesas and Line Islands.

GUTTATA Gmelin, 1791   [32-70mm]   **193**
Two weak subspecies:
guttata guttata: Solid; strongly spotted dorsal pattern and dark blotches on both lips; callus on left margin strong, usually white; E. Indian O. to Melanesia.
guttata azumai: Thinner and more fragile; dorsal pattern usually weak and poorly developed, often no dark blotch on outer lip; callus weak or absent; N. China Sea.

HADDNIGHTAE Trenberth, 1973 [20-35mm] **197**
No variants recognized.
HAMMONDAE Iredale, 1939 [9-18mm] **66**
Three weak variants:
hammondae var. *typical*: Rather large; terminals usually dark; dorsum often minutely flecked and not distinctly zonate; base white; E. Australia.
hammondae var. *dampierensis*: Smaller; terminals usually dark; dorsum zonate as well as flecked; base pale tan; N.W. Australia.
hammondae var. *raysummersi*: Small; terminals often pale; dorsum zonate, flecking usually weak; base white; Philippines to Solomons.

HELVOLA Linnaeus, 1758 [8-36mm] **251**
Very variable in coloration. No variants recognized.
Familiar synonyms: *argella*—E. Africa; *callista*—Oceania; *citrinicolor*—W. Australia; *gereti*—Polynesia; *hawaiiensis*—Hawaii; *mascarena*—N. & C. Indian O.; *meridionalis*—S. Africa.
HIRASEI Roberts, 1913 [40-61mm] **87**
Two subspecies recognized:
hirasei hirasei: Moderately inflated; marginal spots visible in dorsal view; terminals not strongly blotched; base whitish; Japan to Philippines.
hirasei queenslandica: Greatly inflated; marginal spots not visible in dorsal view; terminals distinctly blotched with dark; base tinged with yellow; N.E. Australia.
HIRUNDO Linnaeus, 1758 [8-24mm] **83**
No variants recognized.
Familiar synonyms: *cameroni, endela*—E. Australia; *francisca*—W. & C. Indian O.; *neglecta*—Andamans to Japan & New Guinea; *rouxi*—Melanesia & Micronesia.

HISTRIO Gmelin, 1791 [23-88mm] **161**
No variants recognized.
Familiar synonym: *westralis*—N. Australia.
HUMPHREYSII Gray, 1825 [10-26mm] **92**
No variants recognized.
Familiar synonym: *yaloka*—Fiji.
HUNGERFORDI Sowerby, 1888 [22-40mm] **203**
No variants recognized.
Familiar synonym: *coucomi*—E. Australia.
INTERRUPTA Gray, 1824 [14-27mm] **89**
No variants recognized.
IRRORATA Gray, 1828 [8-17mm] **77**
No variants recognized.
ISABELLA Linnaeus, 1758 [11-54mm] **198**
Two subspecies:
isabella isabella: Terminals shorter and less pointed; adults without a brownish basal ring; Indo-Pacific.
Two weak and broadly intergrading variants:
i.i. var. *typical*: Orange terminal spots separated or confluent, seldom containing a dark spot within.
i.i. var. *controversa*: Orange terminal spots confluent and containing large dark brown spots within; best developed in Hawaiian Islands.
Familiar synonyms: *atriceps*—E. Oceania; *lekalekana*—C. & W. Pacific; *lemuriana*—E. Indian O.; *rumphii*—Andaman Sea.

isabella isabellamexicana **(97):** Terminals somewhat longer and more pointed; adults with deep buffy brown ring below the margins; Pacific Central America.
Familiar synonym: *mexicana*.
KATSUAE Kuroda, 1960 . [19-23mm] **240**
Two variants:
katsuae var. *typical*: Fossula reduced or absent; teeth stained with brown; basal spotting fine.
katsuae var. *musumea*: Fossula well developed; teeth white, unstained; basal spotting larger.
KIENERI Hidalgo, 1906 [8-24mm] **131**
Two variants:
kieneri var. *typical*: Posterior columellar teeth only gradually becoming longer than the anterior teeth; most distinct in W. & C. Indian O.
Familiar synonym: *reductesignata*.
kieneri var. *depriesteri*: Posterior columellar teeth distinctly and suddenly longer than the anterior; most distinct in E. Indian O. & Pacific O.
Familiar synonyms: *landeri; schneideri*—Melanesia; *vitiensis*.
KINGAE Rehder & Wilson, 1975 [14-22m] **133**
No variants recognized.
LABROLINEATA Gaskoin, 1849 [8-31mm] **151**
No variants recognized.
Familiar synonyms: *flaveola; helenae*—C. Melanesia; *nashi*—E. Australia.
LAMARCKII Gray, 1825 [18-51mm] **209**
Two variants:
lamarckii var. *typical*: Numerous ocellated gray spots dorsally; fossula tends to have 3-5 ribs; mostly E. Africa & C. Indian O.
lamarckii var. *redimita*: Dorsal spots plain white, gray spots restricted to lower sides; fossula with 1-3 ribs; mostly E. Indian Ocean.
Slender and broad forms also occur.
LANGFORDI Kuroda, 1938 [41-58mm] **206**
No variants recognized.
Familiar synonym: *moretonensis*—E. Australia.
LENTIGINOSA Gray, 1825 [17-38mm] **139**
Two weak subspecies:
lentiginosa lentiginosa: Oval; dorsum heavily marked with brown spots and flecks; base whitish to yellowish; terminals not very distinct; Persian Gulf to India.
Two varieties:
l.l. var. *typical*: Dorsal banding absent or indistinct.
l.l. var. *buhariensis*: Dorsum with three distinct brownish bands or rows of blotches; S. India.
lentiginosa dancalica: More cylindrical; dorsum lightly spotted and with one to three darker bands or zones; base white; terminals distinct; S. Red Sea.
LEUCODON Broderip, 1828 [69-83mm] **208**
The few known specimens have been split into two subspecies:
leucodon leucodon: Columellar teeth finer, more numerous than labial teeth; base tan, teeth white; dorsal pattern usually complete; Indian O.
leucodon angioyna: Columellar teeth heavier, equal to or fewer than labial teeth; base and teeth both pale; dorsal pattern often lacking in middorsal area; W. Pacific O.
LIMACINA Lamarck, 1810 [12-37mm] **155**
No variants recognized. Note that juveniles with light colors and smooth shells may be quite large in size and that pustules are quite variable in adults.

Familiar synonyms: *facifer*—Melanesia; *interstincta*—W. & C. Indian O.; *monstrans*—E. Australia.

LISETAE Kilburn, 1975   [13-17mm]   253
No variants recognized. Very close to *midwayensis*.

LUCHUANA Kuroda, 1960   [15-23mm]   218
No variants recognized. Extremely close to *pallidula*.

LURIDA Linnaeus, 1758   [14-61mm]   98
Specimens from mid-Atlantic islands may be smaller and more callused than average specimens from continental localities. No variants recognized.
Familiar synonyms: *minima; oceanica*—Mid-Atl. Islands.

LUTEA Gmelin, 1971   [9-22m]   70
No variants recognized.
Familiar synonym: *bizonata*—N.W. Australia.

LYNX Linnaeus, 1758   [18-85mm]   111
Size and color very variable, and cylindrical and broad forms may occur in any population. No variants recognized.
Familiar synonyms: *caledonica*—E. Australia to Oceania; *pacifica*—Fiji; *vanelli*—Japan to E. Indies; *williamsi*—Red Sea.

MACANDREWI Sowerby, 1870   [9-24mm]   81
No variants recognized.

MACULIFERA Schilder, 1932   [31-89mm]   188
No variants recognized.

MAPPA Linnaeus, 1758   [40-100mm]   124
Exceedingly variable, but three weak and inconstant variants can be recognized:
mappa var. typical: Usually inflated; aperture usually not orange; marginal spotting weak or absent; Japan to Australia.
mappa var. *alga:* Usually more cylindrical; aperture broadly bright orange; marginal spots small, numerous, extending onto base; Indian O.
Familiar synonym: *geographica*—Andaman Sea.
mappa var. *viridis:* Shape variable; aperture narrowly orange; marginal spots large or small, numerous; Pacific Islands.
Familiar synonym: *rewa*—Fiji.
Specimens with reddish tints to the base or entire shell are sometimes found; the base color is often very variable.

MARGARITA Dillwyn, 1817   [10-20mm]   144
No variants recognized.
Familiar synonym: *tricornis*—Indian O.

MARGINALIS Dillwyn, 1827   [16-35mm]   230
No variants recognized.
Familiar synonym: *pseudocellata*—Arabia.

MARGINATA Gaskoin, 1849   [43-70mm]   223
Two strong subspecies:
marginata marginata: Relatively light in weight, elongated; pinkish dorsally with grayish marginal band; dorsum usually heavily blotched with brown; teeth of outer lip short, white; usually over 50mm long; S.W. Australia.
marginata ketyana: Thicker, more oval; white dorsally, no gray marginal band; dorsum sparsely to heavily brown-spotted; teeth of outer lip longer, reaching marginal spots; usually under 50mm long; C.W. Australia.

MARIAE Schilder, 1927   [9-20mm]   220
No variants recognized.

MARICOLA Cate, 1976   [14mm]   257
Doubtful species. Very close to *margarita*.

MARTINI Schepman, 1907 [12-20mm] **194**
No variants recognized.
Familiar synonym: *superstes*—New Hebrides.
MAUIENSIS Burgess, 1967 [10-14mm] **199**
No variants recognized. Perhaps better considered a subspecies of *margarita*.
MAURITIANA Linnaeus, 1758 [43-130mm] **177**
No variants recognized.
Familiar synonyms: *calxequina*—Pacific O.; *regina*—E. Indian O.
MICRODON Gray, 1828 [6-15mm] **91**
No variants recognized.
Familiar synonyms: *chrysalis*—W. Indian O.; *granum*—Melanesia.
MIDWAYENSIS Azuma & Kurohara, 1967 [12-22mm] **215**
No variants recognized.
MILIARIS Gmelin, 1791 [17-59mm] **224**
Two subspecies:
miliaris miliaris: Dorsum yellowish to grayish or bluish, with distinct white spots; Japan to E. Australia.
Familiar synonyms: *differens*—Malaysia; *diversa*—N.E. Australia; *inocellata*, *magistra*—Japan; *metavona*—E. Australia.
miliaris eburnea: Dorsum solid white, rarely with pale yellowish tones; Melanesia.
Familiar synonym: *mara*.
Additionally, *miliaris* and *lamarckii* may integrade near Singapore.
MINORIDENS Melvill, 1901 [6-12mm] **65**
No variants recognized.
Familiar synonym: *blandita*—E. Australia.
MONETA Linnaeus, 1758 [10-40mm] **149**
Highly variable in size, shape, and color, presumably with ecological conditions, but no variants recognized.
Familiar synonyms: *barthelemyi*—C. Pacific to Melanesia; *icterina*—E. Africa; *mercatorium*, *numisma*—Indian O.; *rhomboides*—E. Indian O. to W. Pacific.
MUS Linnaeus, 1758 [30-67mm] **116**
Two variants perhaps only reflecting ecological conditions:
mus typical: Columellar teeth weak or absent except at ends; terminals relatively shorter; shallow water.
Familiar synonym: *bicornis*.
mus deep-water: Columellar teeth well formed; terminals more pointed; deep water.
NEBRITES Melvill, 1888 [15-42mm] **160**
No variants recognized. Probably just a variant of *erosa*.
NIGROPUNCTATA Gray, 1828 [17-41mm] **190**
No variants recognized.
NIVOSA Broderip, 1827 [35-75mm] **172**
Large and small forms occur.
NUCLEUS Linnaeus, 1758 [11-31mm] **205**
No variants recognized.
Familiar synonyms: *gemmosa*—Oceania; *madagascariensis*—W. & C. Indian O.; *sturanyi*—Red Sea & Arabia.
OCELLATA Linnaeus, 1758 [14-57mm] **217**
No variants recognized.
Familiar synonym: *pretiosa*.
ONYX Linnaeus, 1758 [24-57mm] **106**
A variable species, but four subspecies can be defined:

onyx onyx: Brown dorsal zones overlaid by wide tan glaze with white margins; lateral teeth black; columellar teeth weak; aperture wide anteriorly; Japan to New Guinea.

onyx melanesiae: Dorsally tan to reddish tan, the zones not readily visible; lateral teeth reddish; columellar teeth heavy anteriorly; aperture narrower; relatively short and broad in shape; Melanesia. Perhaps the Pacific O. representative of *o. succincta.*

onyx succincta: Dorsal zones distinct or covered with brown glaze lacking white margins; lateral teeth reddish; aperture wide anteriorly; shape relatively elongated; N. & W. Indian O.

Two variants:

o.s. var. typical: Dorsal zones usually visible through glaze or glaze reduced in area; Burma to Persian Gulf.

o.s. var. *adusta:* Dorsal zones in adults totally obscured by brownish glaze; Africa to Seychelles.

Familiar synonym: *persica*—Arabia.

onyx nymphae: Dorsal zones absent or extremely indistinct; whitish to very pale tan above and below, terminals darker; Chagos Islands.

OVUM Gmelin, 1791   [16-41mm]   **138**

Coloration of dorsum and teeth varies considerably, but no variants are recognized. Very close to *errones.*

Familiar synonyms: *bartletti, chrysostoma*—orange teeth, E. Australia & Melanesia; *cruenta; palauensis*—W. Micronesia.

OWENII Sowerby, 1837   [8-27mm]   **88**

Strength and sharpness of the lateral callus varies, as does intensity of dorsal, terminal blotch, and lateral spot color. No variants recognized.

Familiar synonyms: *menkeana; vasta*—S. E. Africa.

PALLIDA Gray, 1824   [17-32mm]   **187**

No variants recognized.

Familiar synonym: *insulicola*—Malaysia.

PALLIDULA Gaskoin, 1849   [11-29mm]   **247**

Lateral spots may be absent or present. No variants recognized.

Familiar synonyms: *rhinoceros*—E. Australia & Melanesia; *simulans*—N.W. Australia.

PANTHERINA Solander, 1786   [37-118mm]   **169**

Dorsal coloration is extremely variable in intensity, but no variants are recognized. Apparent hybrids with *tigris* were described as *catulus.*

Familiar synonym: *funebralis*—very dark.

PIPERITA Gray, 1825   [17-32mm]   **135**

(= *comptoni* of first edition; see Schilder, 1975, *Veliger,* 7(1): 37.) Two major variants, each with thinner and paler varieties:

piperita var. typical: Shell oblong or subcylindrical, callus weak or absent.

piperita var. *comptoni:* Shell oval, callused.

Familiar synonyms: *casta; dissecta; mayi; trenberthae.*

PORARIA Linnaeus, 1758   [10-29mm]   **221**

No variants recognized.

Familiar synonyms: *scarabaeus*—Pacific O.; *theoreta*—E. Australia; *wilhelmina*—W. Australia.

PORTERI Cate, 1966   [47-63mm]   **228**

A slender, rather pale form occurs as well as a more inflated and brightly colored form. No variants recognized.

Familiar synonym: *joycae.*

PULCHELLA Swainson, 1823   [23-48mm]   **162**
There is much variation in adult size and development of dorsal pattern. No variants recognized.
Familiar synonyms: *novaebritanniae*—Melanesia; *pericalles, vayssierei*—Arabia.
PULCHRA Gray, 1824   [21-76mm]   **112**
No variants recognized.
PULICARIA Reeve, 1846   [13-22mm]   **207**
No variants recognized.
PUNCTATA Linnaeus, 1771   [7-22mm]   **73**
Two weak subspecies:
punctata punctata: Averaging over 10mm long and somewhat broader; dorsum rarely with three dark zones; teeth often lined with yellow; Indo-Pacific.
Familiar synonyms: *atomaria*—Andamans to Japan and New Guinea; *carula*—N.W. Australia; *iredalei*—Oceania; *persticta*—E. Australia; *stercusmuscarum*.
punctata trizonata: Smaller, about 10mm long, and narrower; dorsum with three broad pale brown zones visible; teeth not lined with yellow; E. Polynesia.
PYRIFORMIS Gray, 1824   [16-37mm]   **174**
Two weak variants:
pyriformis var. typical: Columellar teeth marked with brown; lateral spots usually distinct; E. Indian O. to Philippines & Australia.
Familiar synonyms: *kaiseri, smithi*—N.Australia.
pyriformis var. *angioyorum:* Columellar teeth unmarked or nearly so; lateral spots weak or absent; S. India.
PYRUM Gmelin, 1791   [17-52mm]   **115**
Three ecological varieties:
pyrum var. typical: Base deep reddish brown or orange; cylindrical; shallow water; Mediterranean & W. Africa.
Familiar synonyms: *angolensis; insularum; maculosa; senegalensis*.
pyrum var. *petitiana* **(158)**: Base reddish tan to creamy white with orange tones; broadly cylindrical; moderately deep water (20 fathoms); W. Africa.
pyrum var. *angelicae:* Base white; broadly pyriform with heavy callus; deep water (50-80 fathoms); W. Africa.
QUADRIMACULATA Gray, 1824   [14-32mm]   **147**
No variants recognized.
Familiar synonyms: *garretti*—C. Melanesia; *thielei*—N. Australia.
RABAULENSIS Schilder, 1964   [20-29mm]   **245**
There is considerable variation in length and strength of the posterior terminals, perhaps a function of age. No variants recognized.
RASHLEIGHANA Melvill, 1888   [11-29mm]   **211**
No variants recognized. Extremely close to *teres* var. *latior* and perhaps just a variant.
REEVEI Sowerby, 1832   [26-45mm]   **126**
Slender and inflated forms are known.
ROBERTSI Hidalgo, 1906   [13-32mm]   **154**
No variants recognized.
ROSSELLI Cotton, 1948   [44-64mm]   **110**
No variants recognized.
SAKURAI Habe, 1970   [40-60mm]   **256**
No variants recognized. Extremely similar to *hirasei* and perhaps merely a variant of that species.
SANGUINOLENTA Gmelin, 1791   [14-28mm]   **85**
No variants recognized.

SAULAE Gaskoin, 1843 [15-29mm] **82**
Highly variable, but two weak subspecies can be recognized:
  saulae saulae: Interstices of teeth usually orangish; terminals often tan or orangish; usually larger; N. Indian O. to Japan and Queensland; Palau.
  Familiar synonyms: *jensostergaardi*—Micronesia; *nugata*—E. Australia; *siasiensis*—Philippines.
  saulae crakei: Interstices of teeth usually white; terminals usually whitish; smaller; N.W. Australia.

SCHILDERORUM Iredale, 1939 [22-43mm] **64**
Two strong subspecies:
  schilderorum schilderorum: Aperture nearly straight posteriorly; aperture and most of base white; S. & C. Pacific Islands.
  Familiar synonym: *arenosa.*
  schilderorum kuroharai: Aperture strongly angled posteriorly; base pale tan, including edges of aperture; China Sea.

SCURRA Gmelin, 1791 [23-57mm] **216**
Two variants:
  scurra var. typical: Lateral spots smaller and more crowded; terminal spots weak and inconspicuous; posterior terminals pinched-in basally; best developed in Indian O.
  scurra var. *indica:* Lateral spots usually larger and less crowded; terminal spots dark and conspicuous; posterior terminals not distinctly pinched-in at base; best developed in Pacific O.
  Familiar synonyms: *antelia*—E. Australia; *retifera*—Oceania; *vono*—Polynesia.

SEMIPLOTA Mighels, 1845 [7-37mm] **212**
A rare ovate and callused variety has been called *annae.* Very close to *limacina* and perhaps best considered a subspecies.

SERRULIFERA Schilder & Schilder, 1938 [6-13mm] **68**
No variants recognized.

SPADICEA Swainson, 1823 [30-81mm] **127**
No variants recognized.

SPURCA Linnaeus, 1758 [12-39mm] **182**
Two strong subspecies:
  spurca spurca: Relatively narrow, almost cylindrical; base toned with pale brown; Mediterranean & W. Africa.
  Familiar synonym: *verdensium*—W. Africa.
  spurca acicularis: Broadly ovate, the callus well developed; base white; W. Atlantic O. and Mid-Atlantic Islands.
  Familiar synonym: *sanctaehelenae*—Mid-Atlantic Islands.

STAPHYLAEA Linnaeus, 1758 [7-28mm] **213**
No variants recognized.
  Familiar synonyms: *consobrina*—Oceania; *descripta*—E. Australia; *laevigata*—W. & C. Indian O.

STERCORARIA Linnaeus, 1758 [26-97mm] **173**
Large and small forms occur, as do very dark specimens.
  Familiar synonym: *rattus.*

STOLIDA Linnaeus, 1758 [15-46mm] **157**
Two variable and weak subspecies:
  stolida stolida: Columellar teeth longer, often toned tan; usually two squarish lateral blotches on each side; tends to be more elongated and darker; E. Africa to Micronesia.
  Familiar synonyms: *crossei*—Oceania; *diauges*—W. Indian O.

stolida brevidentata: Columellar teeth very short, white; lateral blotches absent or very weak; tends to be oval and pale; N. Australia.
Familiar synonyms: *deceptor, fluctuans.*
SUBTERES Weinkauff, 1881 [14-29mm] **109**
No variants recognized.
SUBVIRIDIS Reeve, 1835 [18-63mm] **163**
subviridis subviridis: Dorsum either without a blotch or with irregular blotches and interrupted tan zones; E. Australia & Melanesia.
Familiar synonyms: *anceyi*—New Caledonia; *vaticina*—S.E. Australia.
subviridis dorsalis: Dorsum with a large, well defined dorsal blotch and often a smaller blotch at the lower right side; N.W. Australia & S. Papua.
SULCIDENTATA Gray, 1824 [20-77m] **105**
No variants recognized.
SUMMERSI Schilder, 1958 [12-18mm] **95**
No variants recognized. Uncomfortably close to *pallidula.*
SURINAMENSIS Perry, 1811 [23-48mm] **175**
Both narrow, weakly callused shells and broad, heavily callused shells are known.
No variants recognized.
TALPA Linnaeus, 1758 [23-105mm] **233**
Pale and dark forms occur.
Familiar synonyms: *imperialis*—W. Indian O.; *saturata*—Oceania.
TERAMACHII Kuroda, 1938 [58-78mm] **232**
No variants recognized.
TERES Gmelin, 1791 [7-45mm] **219**
Very variable in shape and pattern, but two variants can be recognized:
teres var. typical: Relatively narrow, often with nearly parallel sides; lateral spots usually small; Indo-Pacific.
Familiar synonyms: *alveolus*—C. & W. Indian O.; *pellucens*—Micronesia; *pentella*—E. Australia; *subfasciata*—Melanesia; *tabescens*—Malaysia.
teres var. *latior:* Strongly inflated, with convex sides; lateral spots usually very large; E. Polynesia.
Familiar synonym: *eunota.*
TESSELLATA Swainson, 1822 [15-55mm] **195**
No variants recognized.
TESTUDINARIA Linnaeus, 1758 [74-144mm] **164**
No variants recognized.
Familiar synonyms: *ingens*—C. & W. Indian O.; *testudinosa*—Oceania.
TEULEREI Cazenavette, 1846 [33-67mm] **239**
No variants recognized.
THOMASI Crosse, 1865 [10-24mm] **96**
(= *ostergaardi* of first edition; see Clover, 1978, *Pariah* # 4.) No variants recognized.
TIGRIS Linnaeus, 1758 [42-153mm] **176**
Dark and light forms common.
Familiar synonyms: *ambolee*—Fiji; *lyncichroa*—Oceania; *pardalis*—W. Pacific; *schilderiana*—Hawaiian Islands.
TURDUS Lamarck, 1810 [16-57mm] **114**
Variable in size, density of spotting, and development of callus, but no variants recognized.
Familiar synonyms: *pardalina*—Gulf of Suez; *winckworthi*—Gulf of Oman to Persian Gulf.

URSELLUS Gmelin, 1791 [6-19mm] **80**
No variants recognized.
Familiar synonyms: *amoeba*—W. Melanesia; *coffea*.
VALENTIA Perry, 1811 [64-98mm] **125**
No variants recognized.
VENTRICULUS Lamarck, 1810 [32-76mm] **122**
Large and small forms occur. No variants recognized.
VENUSTA Sowerby, 1846 [49-86mm] **120**
Three minor variants:
venusta var. typical: Large (over 70mm) shell pale, whitish to yellowish with few or no orange-brown blotches or spots.
venusta var. *episema*: Large (over 70mm); shell pale tan with many often confluent dark brown blotches, especially at margins.
venusta var. *sorrentoensis*: Small (under 65mm); color variable.
Familiar synonym: *catei*—mantle mark present.
VITELLUS Linnaeus, 1758 [20-100mm] **254**
Very variable in size. No variants recognized.
Familiar synonyms: dama—E. Africa; *orcina*—E. Australia; *polynesiae*—Oceania; *sarcodes*—N.W. & C. Indian O.
VREDENBURGI Schilder, 1927 [13-32mm] **78**
No variants recognized.
WALKERI Sowerby, 1832 [14-37mm] **150**
Two variants:
walkeri var. typical: Middorsal zone varies from virtually absent at northern part of range to broad and continuous in south; purple of teeth not usually extending as broad stains over columella and outer lip; E. Indian O. to N. & E. Australia.
Familiar synonyms: *continens, merista*—N. Australia.
walkeri var. *surabajensis*: Middorsal zone usually wide and nearly continuous; columella and outer lip usually broadly stained with purple; Philippines to Borneo.
XANTHODON Sowerby, 1832 [16-36mm] **167**
No variants recognized.
ZEBRA Linnaeus, 1758 [32-126mm] **189**
Small specimens most common at southern edge of range.
Familiar synonym: *dissimilis*—Brazil.
ZICZAC Linnaeus, 1758 [8-26mm] **69**
Very variable in pattern, but not geographically constant. No variants recognized.
Familiar synonyms: *misella*—E. Africa; *signata*—E. Australia; *undata*—N. Indian O.; *vittata*—S.E. Australia to Oceania.
ZONARIA Gmelin, 1791 [15-43mm] **179**
Two weak subspecies:
zonaria zonaria: Relatively oval and heavy; columellar teeth comparatively long and distinct posteriorly; dorsum variously blotched and mottled; W. Africa.
Familiar synonym: *gambiensis*.
zonaria picta (**171**): Narrower, with more parallel sides and lighter in weight; columellar teeth weaker and short posteriorly; usually a large middorsal blotch; Cape Verde Islands.

CHECKLIST AND PRICING GUIDE TO THE COWRIES (1979)

The following prices represent the current price ranges for the species. Variants are listed separately only where a price differential exists. Lower values usually indicate average specimens from the more heavily collected localities; larger, exceptionally attractive, or otherwise especially desirable specimens may fetch more than the prices indicated, as may freaks of various types, including solid blacks (niger). Note that prices of some species are constantly fluctuating and that the more expensive shells usually are sold on an individual basis at what the market will bear. The best indicator of current prices is the price list of a reputable dealer. Neither the author nor publisher offers shells for sale; this listing is purely a guide to relative values and not an offer to buy or sell. B = beach; NP = not priced.

____ achatidea - $40.-65.
____ albuginosa - $5.-15.
    Galapagos - $15.
____ algoensis (B) -$10.25.
____ angustata - $3.-5.
    (Varieties higher)
____ annettae - $.50-1.50
    aequinoctialis - $400.-600.
____ annulus - $.10-.25
    obvelata - $1.-2.
____ arabica - $.50-1.50
    Africa - $1.50-2.50
    grayana - $3.-5.
____ arabicula - $.75-1.50
____ argus - $2.50-4.
____ armeniaca - $1000.-1500.
    hesitata - $5.-7.
    howelli - $35.-60.
    beddomei - $8.-15.
____ artuffeli - $15.-25.
    Beach - $3.-6.
____ asellus - $.35-1.
    Giants - $2.

____ aurantium - $400.-600.
    Slight flaws - $250.-400.
____ barclayi - $1400.
    Dead - $500.-1000.
____ becki - $20.-35.
    Beach - $5.-10.
____ bernardi - $900.-1200.
____ bicolor - $3.-4.50
    (Varieties higher)
____ bistrinotata - $1.50-2.
____ boivini - $.75-1.25
____ bregeriana - $16.-25.
____ broderipii - $2500.-3500.
    Flaws - $500.-1500.
____ camelopardalis - $6.-10.
    Flaws - $2.-5.
____ capensis (B) - $3.-7.
    amphithales (B) - $25.-60.
____ caputdraconis - $3.-5.
____ caputserpentis - $.15-.50
    Odd localities - $1.-1.50
____ carneola - $.50-1.50
    Hawaii - $5.-10.

____ Tuamotus - $6.-10.
____ catholicorum - $5.-10.
____ caurica - $.50-1.50
    Niger - $20.-40.
____ cernica - $15.-30.
    Beach - $4.-10.
    Hawaii - $40.
____ cervinetta - $1.-4.
____ cervus - $7.-10.
    Giants - $20.-30.
____ childreni - $10.-20.
    Beach - $5.-8.
____ chinensis - $.50-1.25
    W. Australia - $10.-15.
    Hawaii - $15.-25.
    coloba - $15.-25.
____ cicercula - $1.25-1.50
    Giants - $2.-3.
____ cinerea - $2.-4.
____ citrina - $75.
    Beach - $6.-15.
____ clandestina - $.50-1.
____ contaminata - $15.-20.
____ coxeni - $1.-3.50
____ cribellum - $15.-35.
    Dead - $7.50-10.
____ cribraria - $1.-2.
    Giants - $2.50-4.
    Niger - $40.-50.
    Melwardi - $20.-25.
    W. Australia - $6.50-10.
____ cruickshanki - $350.-500.
____ cumingii - $45.-75.
    Beach - $10.-20.
    astaryi - $100.-250.
    cleopatra - $300.-350.
____ cylindrica - $1.-1.50
    Niger - $40.
____ dayritiana - $3.-4.
____ decipiens - $6.-8.
    perlae - $125.-200.
    Albinistic - $400.-500.
____ declivis - $10.-20.
____ depressa - $.75-1.50
____ dillwyni - $40.-50.
    Beach - $15.-25.

____ diluculum - $1-3.50
____ edentula (B) - $1.25-2.50
    cohenae - $500.-1000.
____ eglantina -$.75-1.50
    Niger - $30.-35.
____ englerti - $12.-15.
____ erosa - $.25-.75
____ errones - $.25-.75
    Niger - $10.
____ erythraeensis - $7.50-15.
____ esontropia -$25.-30.
    Dead - $6.-15.
____ exusta - $70.-125.
____ felina - $1.-2.
____ fernandoi - NP
____ fimbriata - $.50-1.50
____ friendii - $25.-50.
    vercoi - $100.
    thersites - $20.-45.
    jeaniana - $75.-150.
    contraria - $100.-300.
____ fultoni - $2500.-3500.
____ fuscodentata (B) - $2.-4.50
____ fuscorubra (B) - $20.-50.
____ gangranosa - $3.-5.
____ gaskoini
    10-15mm - $5.-10.
    16-20mm - $15.-20.
____ globulus - $1.-1.50
    Gold giants - $5.-8.
____ goodallii - $35.-40.
    fuscomaculata - $50.-75.
____ gracilis - $.75-1.50
____ granulata - $3-4.50
    cassiaui - $400.-450.
    Beach - $100.-200.
____ guttata - $750.-1200.
    Flaws - $300.-500.
____ haddnightae - $100.-150.
    Beach - $50.
____ hammondae
    Australia - $10.-15.
    raysummersi - $.75-1.
____ helvola - $.50-1.50
____ hirasei - $600.-750.
    Dead - $100.-250.

___ hirundo - $.50-1.50
___ histrio - $1.-2.
    Giants - $4.-5.
___ humphreysi - $2.-5.
___ hungerfordi - $15.-20.
    Australia - $60.-75.
___ interrupta - $3.-5.
___ irrorata - $2.-3.
___ isabella - $.25-1.
    isabellamexicana - $10.-20.
___ katsuae - $500.-600.
    Dead - $50.-100.
___ kieneri - $.75-1.50
___ kingae - NP
___ labrolineata - $.50-.75
___ lamarckii - $.50-1.50
___ langfordi - $900.-1000.
    Dead - $100.-300.
___ lentiginosa - $7.-15.
    dancalica - $15.-20.
___ leucodon - $6000.-7000.
    Flaws - $3000.-5000.
___ limacina - $.50-1.50
___ lisetae - $500.-1000.
___ luchuana - $3.50-7.
___ lurida - $1.50-2.50
    Mid-Atlantic - $10.-15.
___ lutea
    Philippines - $.35-1.
    Elsewhere - $2.-3.50
___ lynx - $.25-1.50
___ macandrewi - $100-150.
___ maculifera - $1.-2.50
___ mappa - $2.50-5.
    Colored base - $4.-10.
    Dwarf - $5.-10.
    Niger - $500.-1000.
___ margarita - $4.-6.
___ marginalis - $15.-25.
___ marginata - $150.-175.
    Flaws - $75.-100.
    ketyana - $200.-250.
___ mariae - $15.-25.
    Beach - $8.-10.
___ maricola - NP
___ martini - $45.-75.

    Beach - $10.-15.
___ mauiensis - $30.-35.
___ mauritiana - $1.-2.
    Niger - $150.-200.
___ microdon - $2.-3.
___ midwayensis - $600.-1000.
    Dead - $200.-300.
___ miliaris - $.75-1.25
    eburnea - $3.50-6.
___ minoridens - $1.-1.50
___ moneta - $.25-.50
___ mus - $10.-20.
    Tuberculate - $45.-60.
    Deep-water - NP
___ nebrites - $1.-2.
___ nigropunctata - $7.50-10.
___ nivosa
    Under 50mm - $15.-25.
    Over 50mm - $45.-50.
___ nucleus - $.75-1.50
___ ocellata - $.75-2.
___ onyx - $.75-1.50
    melanesiae - $12.-15.
    succincta - $2.-4.
    adusta - $1.50-2.
    nymphae - $300.-350.
___ ovum - $.75-1.50
___ owenii - $30.-40.
    Beach - $10.
___ pallida - $1.50-3.50
___ pallidula - $.75-1.50
___ pantherina - $1.50-2.
    Colors - $3.-9.
___ piperita - $3.-5.
    (Varieties higher)
___ poraria - $.50-1.
___ porteri - $1700.-2000.
    Dead - $200.-400.
___ pulchella
    Strong pattern - $12.-15.
    Weak pattern - $3.50-7.50
___ pulchra - $7.50-12.
___ pulicaria - $6.-7.
___ punctata - $.75-2.50
    trizonata - $6.-7.50
___ pyriformis - $1.-2.

    angioyorum - $10.
___ pyrum - $1.50-2.50
    petitiana - $40.-75.
    angelicae - $200.-400.
___ quadrimaculata - $1.-3.
___ rabaulensis
    Philippines - $125.-150.
    Elsewhere - $200.-250.
___ rashleighana - $75.-100.
___ reevei - $25.-50.
___ robertsi - $.75-1.
___ rosselli - $350.-450.
    Flaws - $200.-250.
___ sakurai - $600.-1000.
    Dead - $100.-200.
___ sanguinolenta - $40.-75.
___ saulae - $20.-25.
___ schilderorum - $3.-6.
    kuroharai - $400.-500.
    Dead - $100.-200.
___ scurra - $2.-3.
___ semiplota - $25.-60.
___ serrulifera - $25.
___ spadicea - $1.25-2.
    Albino - $50.
___ spurca - $1.50-3.
    Mid-Atlantic - $10.-15.
___ staphylaea - $.35-1.
___ stercoraria - $6.-9.
___ stolida - $5.-10.
___ subteres - $35.-40.
___ subviridis - $1.-2.
___ sulcidentata - $4.50-7.50
___ summersi - $35.-45.

___ surinamensis - $500.
    Dead - $250.-350.
___ talpa - $1.-1.50
___ teramachii - $1000.-1500.
    Dead - $200.-500.
___ teres - $.50-1.50
    latior - $40.-75.
___ tessellata - $25.-50.
___ testudinaria - $4.50-7.50
___ teulerei - $5.-10.
___ thomasi - $600.-800.
    Beach - $90.-120.
___ tigris - $.75-1.50
    Hawaii, 100m-$30.
        125mm-$200.
___ turdus - $1.-3.
___ ursellus - $2.50-5.
___ valentia - $1000.-1750.
    Dead - $250.-600.
___ ventriculus - $7.-10.
___ venusta
    Over 70mm - $40.-75.
    Under 65mm - $20.-40.
___ vitellus - $.50-1.50
___ vredenburgi - $150.-250.
    Beach - $40.-100.
___ walkeri - $1-3.50
___ xanthodon - $.75-1.50
___ zebra - $5.-10.
___ ziczac - $.75-1.
    Africa - $1.50-3.50
___ zonaria - $3.50-7.50
    picta - $25.-45.

SYNONYMIC INDEX

The following index attempts to present valid names applied to living cowries between 1758 and the end of 1978. It has proved impractical to associate many names with the subspecies and varieties recognized in the *Synopsis,* so the simple expedient of referring all names only to species is resorted to. The synonymy is mainly based on Schilder & Schilder, 1971, *Inst. Roy. Sci. Nat. Belgique,* 85, but most modern names have been reviewed and many changes are evident. See the *Synopsis* for the status of some subspecies and major varieties.

abbreviata Dautzenberg 1903 = hirundo
aberrans Ancey 1882 = clandestina
ACHATIDEA Sowerby 1837 = valid, 191
achatina Perry 1811 = ventriculus
acicularis Gmelin 1791 = spurca
adansonii Blainville 1825 = mauritiana
adelinae Roberts 1885 = goodalli
adonis Rous 1905 = carneola
adusta Lamarck 1810 = onyx
aequinoctialis Schilder 1933 = annettae
affinis Gmelin 1791 = globulus
alauda Menke 1830 = tigris
alba Blainville 1826 = zonaria
alba Coen 1949 = fuscodentata
alba Cox 1879 = armeniaca
alba Sowerby 1832 = turdus
alba Sullioti 1924 = gangranosa, punctata, schilderorum
albata Beddome 1898 = angustata
albella Lamarck 1822 = caputserpentis
albicilla Sullioti 1924 = staphylaea
albida Dautzenberg 1893 = errones
albida Gmelin 1791 = gangranosa
albida Sullioti 1924 = lurida
albida Monterosato 1897 = spurca
albiflora Petrbok 1932 = mauritiana
albina Sullioti 1924 = arabica
albinella Melvill & Standen 1895 = poraria
alboguttata Schroeter 1804 = tigris
albolineata Turton 1932 = capensis
albonitens Melvill 1888 = pantherina
albopunctata Fischer 1807 = erosa
albosignata Coen 1949 = caputserpentis
ALBUGINOSA Gray 1825 = valid, 134
alfredensis Schilder & Schilder 1929 = edentula
alga Perry 1811 = mappa
ALGOENSIS Gray 1825 = valid, 156
alleni Ostergaard 1950 = thomasi
alveolus Tapperone 1882 = teres
amabilis Jousseaume 1881 = walkeri

amarata Moerch 1852 = scurra
ambigua Gmelin 1791 = tigris
ambolee Steadman & Cotton 1943 = tigris
amethystea Linnaeus 1758 = histrio
amethystina Kobelt 1906 = lurida
amiges Melvill & Standen 1915 = chinensis
amoeba Schilder & Schilder 1938 = ursellus
amoena Schilder 1927 = boivinii
amphithales Melvill 1888 = capensis x edentula
anceyi Vayssiere 1905 = subviridis
angelicae Clover 1974 = pyrum
angioyorum Biraghi 1978 = pyriformis
angioyna Raybaudi 1979 = leucodon
angolensis Odhner 1923 = pyrum
ANGUSTATA Gmelin 1791 = valid, 141
angustata Gray 1824 = cervinetta
annae Roberts 1869 = semiplota
ANNETTAE Dall 1909 = valid, 246
annosa Vayssiere 1905 = gangranosa
annularis Perry 1811 = annulus
annulata Donovan 1820 = annulus
annulata Gray 1828 = mariae
annulifera Conrad 1866 = annulus
annulifera Coen 1949 = moneta
ANNULUS Linnaeus 1758 = valid, 136
antelia Iredale 1939 = scurra
aphrodite Rous 1905 = helvola
ARABICA Linnaeus 1758 = valid, 184
ARABICULA Lamarck 1810 = valid, 166
arenosa Dillwyn 1823 = turdus
arenosa Gray 1824 = schilderorum
argella Melvill 1888 = helvola
argentata Dautzenberg & Bouge 1933 = caputserpentis
argiolus Roeding 1798 = scurra
ARGUS Linnaeus 1758 = valid, 113
argusculus Schroeter 1804 = helvola
arlequina Moerch 1852 = histrio
ARMENIACA Verco 1912 = valid, 108
ARTUFFELI Jousseaume 1876 = valid, 119

ASELLUS Linnaeus 1758 = valid, 102
asiatica Schilder & Schilder 1939 = arabica
astaryi Schilder 1971 = cumingii
atlantica Monterosato 1897 = spurca
atomaria Gmelin 1791 = punctata
atrata Gray 1825 = staphylaea
atrata Sullioti 1924 = caurica, decipiens, friendii, stercoraria
atra Dautzenberg 1903 = arabica, mauritiana
atriceps Schilder & Schilder 1938 = isabella
aubryana Jousseaume 1869 = surinamensis
aurantia Coen 1949 = pyrum
AURANTIUM Gmelin 1791 = valid, 117
aurea Shaw 1909 = moneta
aurea Coen 1949 = carneola, eglantina
aureola Schroeter = carneola
auricoma Crosse 1896 = achatidea
auricomata Sullioti 1924 = pantherina
aurora Lamarck 1810 = aurantium
aurora Monterosato 1897 = lurida
australis Schroeter 1804 = helvola
autumnalis Perry 1811 = mus
avalitensis Jousseaume 1894 = erythraeensis
azumai Schilder 1960 = guttata
azurea Schilder 1968 = errones

badia Coen 1949 = lurida
badia Gmelin 1791 = helvola
badionitens Melvill 1888 = pantherina
bakeri Gatliff 1916 = venusta
baltheata Dillwyn 1823 = erosa
bandata Perry 1811 = caputserpentis
barbadensis Verrill 1948 = carneola
barbara Kenyon 1902 = bregeriana
BARCLAYI Reeve 1857 = valid, 74
barthelemyi Bernardi 1861 = moneta
bartletti Steadman & Cotton 1946 = ovum
BECKI Gaskoin 1836 = valid, 67
beddomei Schilder 1930 = armeniaca
berinii Dautzenberg 1906 = punctata
BERNARDI Richard 1974 = valid, 225
bernardinae Preston 1907 = lamarckii
bicallosa Gray 1831 = surinamensis
BICOLOR Gaskoin 1849 = valid, 79
bicornis Sowerby 1870 = mus
bifasciata Gmelin 1791 = cervus
bifasciata Monterosato 1897 = pyrum
bimaculata Gray 1824 = errones
BISTRINOTATA Schilder & Schilder 1937 = valid, 76
bitaeniata Geret 1903 = asellus
bizonata Iredale 1935 = lutea
blaesa Iredale 1939 = caurica
blandita Iredale 1939 = fimbriata
BOIVINII Kiener 1843 = valid, 130
borneensis Kenyon 1902 = helvola
bougei Dautzenberg 1906 = asellus
BREGERIANA Crosse 1868 = valid, 248
brevidentata Sowerby 1870 = stolida
brevirostris Schilder & Schilder 1938 = globulus
brevis Smith 1913 = miliaris
britannica Schilder 1927 = moneta
brocktoni Iredale 1930 = guttata
BRODERIPII Sowerby 1832 = valid, 196
brookei Rous 1905 = miliaris
brunnea Gray 1825 = ocellata
brunnescens Cate 1964 = arabica
buhariensis Jonklaas & Nicolay 1977 = lentiginosa
bulgarica Kojumdgieva 1960 = moneta
bullata Roeding 1798 = cinerea
buttoni Oldroyd 1916 = diluculum
bynei Sullioti 1924 = argus

caerulea Fischer 1807 = cylindrica
caerulea Schroeter 1804 = cinerea, ?labrolineata
caerulescens Mollerat 1890 = lurida
cairnsiana Melvill & Standen 1904 = ?chinensis
calcarata Melvill 1884 = annulus
caledonica Crosse 1869 = lynx
caledonica Coen 1949 = caurica
callista Shaw 1909 = helvola
calophthalma Melvill 1888 = ocellata
calxequina Melvill & Standen 1899 = mauritiana
CAMELOPARDALIS Perry 1811 = valid, 231
camelorum Rochebrune 1884 = annulus
cameroni Iredale 1939 = hirundo
canaliculata Roeding 1798 = arabica
cancellata Gmelin 1791 = helvola
candida Pease 1865 = clandestina
candida Dautzenberg & Bouge 1933 = moneta
candida Coen 1949 = pulicaria
candidata Sullioti 1924 = caputserpentis
CAPENSIS Gray 1828 = valid, 178
caputanguis Philippi 1849 = caputserpentis
caputcolubri Kenyon 1898 = caputserpentis
CAPUTDRACONIS Melvill 1888 = valid, 238
caputophidii Schilder 1927 = caputserpentis
CAPUTSERPENTIS Linnaeus 1758 = valid, 243
carmen Smith 1912 = nebrites
CARNEOLA Linnaeus 1758 = valid, 142
carneola Chenu 1845 = ventriculus
carnicolor Moerch 1852 = onyx
carnicolor Preston 1909 = xanthodon
carthaginiensis Roeding 1798 = mus
carula Iredale 1939 = punctata
cassiaui Burgess 1965 = granulata
casta Schilder & Summers 1963 = piperita
castanea Anderson 1836 = angustata
castanea Higgins 1868 = fuscorubra
castanea Roeding 1798 = mauritiana
catei Schilder 1963 = venusta
CATHOLICORUM Schilder & Schilder 1938 = valid, 229
catulus Schilder 1924 = pantherina x tigris
CAURICA Linnaeus 1758 = valid, 99
cavia Steadman & Cotton 1943 = isabella
cerea Sullioti 1924 = nucleus
CERNICA Sowerby 1870 = valid, 234
cervina Lamarck, 1822 = cervus
CERVINETTA Kiener 1843 = valid, 168
CERVUS Linnaeus 1771 = valid, 185
ceylonica Schilder & Schilder 1938 = nebrites
chalcedonia Perry 1811 = helvola
CHILDRENI Gray 1825 = valid, 129
CHINENSIS Gmelin 1791 = valid, 186
chionella Sullioti 1924 = moneta
chionia Melvill 1888 = tigris
chlorizans Melvill 1888 = erosa
cholmondeleyi Melvill 1888 = gracilis
chrysalis Kiener 1843 = microdon
chrysophaea Melvill 1888 = errones
chrysostoma Schilder 1927 = ovum
CICERCULA Linnaeus 1758 = valid, 214
cineracea Sulliotti 1924 = stercoraria
CINEREA Gmelin 1791 = valid, 101
cinerea Bouge 1961 = mauritiana
cinereoviridescens Bouge 1961 = mappa
cinnamonaea Olivi 1792 = pyrum
circumvallata Schilder & Schilder 1933 = moneta
CITRINA Gray 1825 = valid, 255
citrinicolor Iredale 1935 = helvola
CLANDESTINA Linnaeus 1767 = valid, 75
clara Gaskoin 1851 = isabella
cleopatra Schilder & Schilder 1938 = cumingii
coeca Roeding 1798 = poraria

coerulea Nardo 1847 = lurida
coerulea Karsten 1789 = moneta
coerulea Perry 1811 = annulus
coerulescens Schroeter 1804 = errones
coffea Sowerby 1870 = ursellus
cohenae Burgess 1965 = edentula x algoensis
coloba Melvill 1888 = chinensis
colorata Dautzenberg 1932 = chinensis
comma Perry 1811 = cribraria
commixta Wood 1828 = interrupta
compressa Coen 1949 = pyrum
compressa Dautzenberg 1903 = errones
compta Pease 1860 = cumingii
comptoni Gray 1847 = piperita
concatenata Dautzenberg 1903 = argus
concava Sowerby 1870 = caurica
concolor Kobelt 1906 = lurida
confusa Coen 1949 = pyrum
consobrina Garret 1879 = staphylaea
conspurcata Gmelin 1791 = stercoraria
CONTAMINATA Sowerby 1832 = valid, 93
continens Iredale 1935 = walkeri
contraria Iredale 1935 = friendii
contrastriata Perry 1811 = argus
controversa Gray 1824 = isabella
controversa Menke 1829 = gangranosa
convergens Dautzenberg 1932 = chinensis
cornus Roeding 1798 = cylindrica
coronata Schilder 1930 = fuscodentata
corrosa Schilder & Schilder 1938 = caurica
coucomi Schilder 1964 = hungerfordi
couturieri Vayssiere 1905 = eglantina
COXENI Cox 1873 = valid, 153
coxi Brazier 1872 = errones
crakei Cate 1968 = saulae
crassa Gmelin, 1791 = carneola
crenata Roeding 1798 = chinensis
CRIBELLUM Gaskoin 1849 = valid, 242
CRIBRARIA Linnaeus 1758 = valid, 201
crossei Marie 1869 = stolida
cruenta Gmelin 1791 = errones
cruenta Coen 1949 = pyrum
cruentata Roeding 1798 = lynx
CRUICKSHANKI Kilburn 1972 = valid, 202
CUMINGII Sowerby 1832 = valid, 143
curta Mollerat 1890 = lurida
CYLINDRICA Born 1778 = valid, 152
cylindrica Blainville 1826 = teres
cylindrica Mollerat 1890 = lurida
cylindroidea Coen 1949 = isabella
cypraea Linnaeus 1758 = spurca

dama Perry 1811 = vitellus
dampierensis Schilder & Cernohorsky 1965 = hammondae
dancalica Jonklaas & Nicolay 1977 = lentiginosa
dautzenbergi Hidalgo 1907 = goodalli
DAYRITIANA Cate 1963 = valid, 118
deceptor Iredale 1935 = stolida
DECIPIENS Smith 1880 = valid, 165
DECLIVIS Sowerby 1870 = valid, 145
decolorata Dautzenberg 1903 = ziczac
decolorata Gray 1824 = punctata
depravata Dautzenberg 1903 = staphylaea
DEPRESSA Gray 1824 = valid, 181
depriesteri Schilder 1933 = kieneri
derosa Gmelin 1791 = caurica
derosa Risso 1826 = helvola
descripta Iredale 1935 = staphylaea
devians Sulliotti 1924 = annulus
diaphana Coen 1949 = erosa
diaphana Sulliotti 1924 = semiplota

diaugues Melvill 1888 = stolida
differens Schilder 1927 = miliaris
dilacerata Schilder & Schilder 1939 = arabica
dilatata Coen 1949 = arabica
dilatata Monterosato 1897 = spurca
DILLWYNI Schilder 1922 = valid, 84
DILUCULUM Reeve 1845 = valid 183
diluta Monterosato 1897 = pyrum
dispersa Schilder & Schilder 1939 = depressa
dissecta Iredale 1931 = piperita
dissimilis Schilder 1924 = zebra
distans Schilder & Schilder 1938 = contaminata
distinguenda Schilder 1927 = turdus
distorta Iredale 1935 = vitellus
diversa Kenyon 1902 = miliaris
dorsalis Schilder & Schilder 1938 = subviridis
dorsoalbida Williams 1914 = caputserpentis
dracaena Born 1778 = caurica
draco Roeding 1798 = stolida
dranga Iredale 1939 = annulus
dubia Gmelin 1791 = zebra
dubia Gray 1831 = ?caurica
duploreticulata Coen. 1949 = histrio
durbanensis Schilder & Schilder 1938 = fimbriata

ebur Coen 1949 = semiplota
eburnea Barnes 1824 = miliaris
EDENTULA Gray 1825 = valid, 148
efasciata Pallary 1900 = lurida, achatidea
efasciata Monterosato 1897 = lurida
effosa Schilder 1937 = miliaris
EGLANTINA Duclos 1833 = valid, 250
elaoides Melvill 1888 = teres
elizabethensis Rous 1905 = capensis
elliptica Gray 1825 = spurca
elongata Coen 1949 = pyrum
elongata Dautzenberg & Fischer 1906 = spurca
elongata Pallary 1900 = achatidea
elongata Perry 1811 = caurica
emaculata Coen 1949 = cylindrica
emblema Iredale 1931 = angustata
endela Iredale 1939 = hirundo
endua Steadman & Cotton 1943 = moneta
ENGLERTI Summers & Burgess 1965 = valid, 227
episema Iredale 1939 = venusta
epunctata Coen 1949 = diluculum
EROSA Linnaeus 1758 = valid, 200
ERRONES Linnaeus 1758 = valid, 104
erua Steadman & Cotton 1943 = moneta
ERYTHRAEENSIS Sowerby 1837 = valid, 72
ESONTROPIA Duclos 1833 = valid, 204
ethnographica Rochebrune 1884 = moneta
etolu Steadman & Cotton 1943 = moneta
euclia Steadman & Cotton 1946 = bicolor
eugeniae Cate 1975 = xanthodon
eunota Taylor 1916 = teres
exanthema Linnaeus 1767 = zebra
exanthemata Perry 1811 = zebra
exmouthensis Melvill 1888 = cribraria
expallescens Dautzenberg & Bouge 1933 = staphylaea
extrema Iredale 1939 = clandestina
EXUSTA Sowerby 1832 = valid, 107

fabula Kiener 1843 = felina
facifer Iredale 1935 = limacina
faedata Sulliotti 1911 = turdus
fallax Smith 1881 = cribraria
fasciata Gmelin 1791 = stercoraria
fasciata Perry 1811 = erosa

fasciata Donovan 1820 = diluculum
fasciomaculata Coen 1949 = nebrites
FELINA Gmelin, 1791 = valid, 226
feminea Gmelin 1791 = tigris
fergusoni Rous 1905 = vitellus
FERNANDOI Cate 1969 = ?valid, 257
ferruginea Fischer 1807 = lynx
ferruginosa Gmelin 1791 = mus, ?gangranosa
ferruginosa Kiener 1843 = annettae
FIMBRIATA Gmelin 1791 = valid, 241
fimbriatula Sowerby 1870 = semiplota
fimbriolata Carpenter 1872 = albuginosa
fischeri Vayssiere 1910 = gaskoini
flammea Gmelin 1791 = tigris
flava Sulliotti 1922 = cinerea
flava Menke 1829 = moneta
flaveola Linnaeus 1758 = ?labrolineata
flavescens Gray 1824 = asellus
flavescens Schroeter 1804 = chinensis
flavida Dautzenberg 1893 = tigris
flavida Monterosato 1897 = spurca
flavonitens Melvill 1888 = tigris
fluctuans Iredale 1935 = stolida
formosa Gray 1824 = cylindrica
fortis Coen 1949 = staphylaea
fragiliodes Hidalgo 1907 = cinerea
fragilis Linnaeus 1758 = arabica
francisca Schilder & Schilder 1938 = hirundo
FRIENDII Gray 1831 = valid, 192
fuliginosa Perry 1811 = mauritiana
fuliginosa Roeding 1798 = mus
FULTONI Sowerby 1903 = valid, 103
fulva Gmelin, 1791 = pyrum
fulva Gray 1824 = cinerea, isabella
fulva Gray 1828 = onyx
fulva Mollerat 1890 = lurida
fulva Rous 1905 = isabella
funebralis Sulliotti 1924 = pantherina
fusca Coen 1949 = asellus, errones
fusca Monterosato 1897 = pyrum
fuscoapicata Coen 1949 = tigris
FUSCODENTATA Gray 1825 = valid, 170
fuscofasciata Schroeter 1804 = caurica
fuscomaculata Pease 1865 = goodalli
FUSCORUBRA Shaw 1909 = valid, 100
fuscotecta Sulliotti 1924 = mus

gabrielana Gatliff 1929 = miliaris
gabrieli Gatliff 1916 = miliaris
galactina Sulliotti 1924 = capensis
galbula Taylor 1916 = erosa
gambiensis Shaw 1909 = zonaria
GANGRANOSA Dillwyn 1817 = valid, 159
garretti Schilder & Schilder 1938 = quadrimaculata
GASKOINI Reeve 1846 = valid, 242
gedlingae Cate 1968 = carneola
gelasima Melvill 1888 = stolida
gemmosa Perry 1811 = nucleus
gemmula Weinkauff 1881 = arabicula
geographica Schilder & Schilder 1933 = mappa
gereti Vayssiere 1910 = helvola
gibba Coen 1949 = arabica
gibba Gmelin 1791 = stercoraria
gibbosa Coen 1949 = vitellus
gibbosa Gray 1824 = felina
gibbosa Schroeter 1804 = moneta
gillei Jousseaume 1893 = depressa
gilva Sulliotti 1924 = pallida
glauca Roeding 1798 = zebra
globosa Dautzenberg 1903 = lynx
globosa Pallary 1900 = achatidea, pyrum

globosa Vayssiere 1910 = angustata
GLOBULUS Linnaeus 1758 = valid, 86
gloriosa Shikama 1971 = fuscodentata
gondwanalandensis Burgess 1970 = fuscorubra
GOODALLI Sowerby 1832 = valid, 71
GRACILIS Gaskoin, 1849 = valid, 123
grandis Monterosato 1897 = pyrum
GRANULATA Pease 1862 = valid, 140
granulosa Schilder & Schilder 1940 = nucleus
granum Schilder & Schilder 1938 = microdon
grayana Schilder 1930 = arabica
grayi Kiener 1843 = achatidea
greegori Ford 1893 = chinensis
grisea Coen 1949 = staphylaea
GUTTATA Gmelin 1791 = valid, 193
guttata Lamarck 1810 = pantherina
guttata Roberts 1870 = tigris

HADDNIGHTAE Trenberth 1973 = valid, 197
haematites Nardo 1847 = pyrum
halmaja Melvill, 1888 = carneola
HAMMONDAE Iredale, 1939 = valid, 66
hamyi Rochebrune 1884 = turdus
harmondiana Rochebrune 1884 = annulus
harrisi Iredale 1939 = moneta
hartsmithi Schilder 1967 = ?bicolor
hawaiiensis Melvill 1888 = helvola
helenae Roberts 1869 = labrolineata
HELVOLA Linnaeus 1758 = valid, 251
hepatica Coen 1949 = pyrum
hermanni Iredale 1939 = teres
hesitata Iredale 1916 = armeniaca
hesperina Schilder & Summers 1963 = coxeni
hidalgoi Shaw 1909 = teulerei
hilda Iredale 1939 = gracilis
hinnulea Melvill 1888 = tigris
HIRASEI Roberts 1913 = valid, 87
HIRUNDO Linnaeus 1758 = valid, 83
HISTRIO Gmelin 1791 = valid, 161
honoluluensis Melvill 1888 = granulata
howelli Iredale 1931 = armeniaca
HUMPHREYSI Gray 1825 = valid, 92
HUNGERFORDI Sowerby 1888 = valid, 203

icterina Lamarck 1810 = moneta
immaculata Coen 1949 = caurica, helvola
immanis Schilder & Schilder 1939 = arabica
imperialis Schilder & Schilder 1938 = talpa
inaequipartita Monterosato 1897 = spurca
incana Sulliotti 1924 = tigris
incrassata Dautzenberg 1929 = lynx
incrassata Coen 1949 = lurida
indica Gmelin 1791 = scurra
induta Sulliotti 1924 = arabica
inflata Coen 1949 = spurca
ingens Schilder & Schilder 1938 = testudinaria
ingloria Crosse 1878 = surinamensis
inocellata Gray 1825 = miliaris, erosa
inocellata Coen 1949 = lamarckii
inopinata Schilder 1930 = achatidea
insignis Dautzenberg 1903 = poraria
inspersa Link 1807 = caurica
insularum Schilder 1928 = pyrum
insulicola Schilder & Schilder 1938 = pallida
intermedia Gray 1824 = arabica
intermedia Smith 1913 = miliaris
interpunctata Henn 1896 = gracilis
INTERRUPTA Gray 1824 = valid, 89
interstincta Wood 1828 = limacina
inversa Pallary 1900 = spurca
ionthodes Melvill 1888 = tigris
iredalei Schilder & Schilder 1938 = punctata

irescens Sowerby 1870 = gracilis
irina Kiener 1843 = nigropunctata
irvineanae Cox 1889 = stolida
IRRORATA Gray 1828 = valid, 77
ISABELLA Linnaeus 1758 = valid, 198
isabellamexicana Stearns 1893 = isabella
isabelloides Schilder 1924 = isabella
isomeres Iredale 1939 = moneta
iutsui Shikama 1974 = fuscorubra

japonica Schilder 1931 = gracilis
javana Coen 1949 = lynx
jeaniana Cate 1968 = friendii
jenningsia Perry 1811 = limacina
jennisoni Steadman & Cotton 1943 = globulus
jensostergaardi Ingram 1939 = saulae
jousseaumei Vayssiere 1905 = cervus
joycae Clover 1970 = porteri
juvenca Melvill 1888 = pantherina

kaiseri Kenyon 1897 = pyriformis
kalavo Steadman & Cotton 1943 = errones
kaolinica Vredenburg 1919 = erosa
katha Iredale 1939 = microdon
KATSUAE Kuroda 1960 = valid, 240
kauaiensis Melvill 1888 = poraria
kauilani Kenyon 1900 = ?erosa
kawakawa Steadman & Cotton 1943 = asellus
keelingensis Schilder & Schilder 1940 = bistrinotata
kenyonae Schilder & Schilder 1938 = caputserpentis
kermadecensis Powell 1958 = cernica
kesata Steadman & Cotton 1943 = subviridis
ketyana Raybaudi 1978 = marginata
KIENERI Hidalgo 1906 = valid, 131
kiiensis Roberts 1913 = hungerfordi
KINGAE Rehder & Wilson 1975 = valid, 133
korolevu Steadman & Cotton 1943 = hirundo
kunthii Audouin 1827 = lurida
kuroharai Kuroda & Habe 1961 = schilderorum

LABROLINEATA Gaskoin 1849 = valid, 151
lactea Coen 1949 = staphylaea
lactea Wood 1828 = miliaris
lactescens Dautzenberg & Bouge 1933 = erosa
laevigata Dautzenberg 1932 = staphylaea
LAMARCKII Gray 1825 = valid, 209
laminata Sulliotti 1924 = tigris
landeri Schilder & Griffiths 1962 = kieneri
LANGFORDI Kuroda 1938 = valid, 206
latefasciata Schilder 1930 = asellus
laticolor Pallary 1900 = pyrum
latior Melvill 1888 = teres
lekalekana Ladd 1934 = isabella
lemuriana Steadman & Cotton 1946 = isabella
lemurica Schilder & Schilder 1938 = childreni
lenella Iredale 1939 = cylindrica
LENTIGINOSA Gray 1825 = valid, 139
lentiginosa Coen 1949 = declivis
lentigo Roeding 1798 = caurica
leopardus Link 1807 = pantherina
leucochroa Sulliotti 1924 = ?piperita
LEUCODON Broderip 1828 = valid, 208
leucogaster Gmelin 1791 = lurida
leucopis Shaw 1803 = zebra
leucostoma Gaskoin 1843 = teulerei
leucostoma Gmelin 1791 = lynx
leviathan Schilder & Schilder 1937 = carneola
liburnica Coen 1949 = lurida
lienardi Jousseaume 1874 = cicercula
ligata Roeding 1798 = stercoraria

ligata Schroeter 1804 = caurica
lilacina Weinkauff 1881 = limacina
lima Roeding 1798 = nucleus
LIMACINA Lamarck 1810 = valid, 155
limitaris Monterosato 1897 = spurca
limpida Melvill 1888 = carneola
limpida Sulliotti 1922 = spurca
lineata Gmelin 1791 = felina
lineata Kenyon 1902 = tigris
LISETAE Kilburn 1975 = valid, 253
listeri Gray 1824 = felina
listeri Gray 1825 = marginalis
literata Time 1822 = isabella
livida Gmelin 1791 = unrecognizable
loebbeckeana Weinkauff 1881 = carneola
longinqua Schilder & Schilder 1938 = achatidea
longior Iredale 1935 = caurica
lota Linnaeus 1758 = spurca
LUCHUANA Kuroda 1960 = valid, 218
lucida Taylor 1916 = erosa
luctuosa Dautzenberg 1903 = eglantina
lunata Fischer 1807 = spurca
LURIDA Linnaeus 1758 = valid, 98
luridoidea Pallary 1900 = spurca
LUTEA Gmelin 1791 = valid, 70
lyncichroa Melvill 1888 = tigris
LYNX Linnaeus 1758 = valid, 111

MACANDREWI Sowerby 1870 = valid, 81
maccullochi Iredale 1939 = labrolineata
macula Angas 1867 = gracilis
maculata Barnes 1824 = maculifera
maculata Gray 1824 = punctata, zonaria
maculata Perry 1811 = angustata
maculatus Donovan 1820 = cribraria
MACULIFERA Schilder 1932 = valid, 188
maculosa Gmelin 1791 = pyrum
madagascariensis Gmelin 1791 = nucleus
magerrones Iredale 1939 = errones
magistra Melvill 1888 = miliaris
magnifica Coen 1949 = diluculum
major Dautzenberg 1903 = errones
major Dautzenberg 1929 = carneola
major Pallary 1900 = achatidea, lurida, spurca
major Mollerat 1890 = achatidea
malaysiae Schilder & Schilder 1940 = contaminata
MAPPA Linnaeus 1758 = valid, 124
mara Iredale 1939 = miliaris
marcia Iredale 1939 = kieneri
MARGARITA Dillwyn 1817 = valid, 144
margarita Gray 1828 = dillwyni
MARGINALIS Dillwyn 1827 = valid, 230
MARGINATA Gaskoin 1849 = valid, 223
marginata Kiesenwetter 1872 = moneta
marginata Coen 1949 = erosa
MARIAE Schilder 1927 = valid, 220
mariae Schilder 1924 = camelopardalis
MARICOLA Cate 1976 = ?valid, 257
marielae Cate 1960 = cernica
marmorata Blainville 1826 = carneola
marmorata Schroeter 1804 = fimbriata
MARTINI Schepman 1907 = valid, 194
marteli Dautzenberg 1903 = clandestina, hirundo
martiniana Anton 1839 = gangranosa
mascarena Melvill 1888 = helvola
massauensis Schilder 1922 = arabicula
maturata Kira 1959 = cernica
MAUIENSIS Burgess 1967 = valid, 199
MAURITIANA Linnaeus 1758 = valid, 177
maxima Dautzenberg 1903 = moneta

maxima Monterosato 1897 = lurida
mayi Beddome 1898 = piperita
media Monterosato 1897 = lurida
mediocris Schilder & Schilder 1938 = bistrinotata
melanesiae Schilder & Schilder 1937 = *onyx
melanosema Melvill 1888 = gangranosa
melanostoma Sowerby 1825 = camelopardalis
meleagris Roeding 1798 = cervus
melvilli Hidalgo 1906 = felina
melwardi Iredale 1930 = cribraria
menkeana Deshayes 1863 = owenii
mercatorium Rochebrune 1884 = moneta
meridionalis Schilder & Schilder 1938 = helvola
merista Iredale 1939 = walkeri
metavona Iredale 1935 = miliaria
mexicana (of authors) = isabella
michaelis Melvill 1905 = lynx
MICRODON Gray 1828 = valid, 91
MIDWAYENSIS Azuma & Kurohara 1967 = valid, 215
mikado Schilder & Schilder 1938 = caputserpentis
MILIARIS Gmelin 1791 = valid, 224
minima Dunker 1853 = lurida
minima Dautzenberg & Bouge 1933 = carneola
minima Monterosato 1897 = spurca
minima Coen 1949 = pyrum
minor Dautzenberg 1893 & 1903 = argus, caputserpentis, carneola, errones, helvola
minor Pallary 1900 = achatidea, lurida, pyrum, spurca
minor Quenstedt 1884 = ursellus
MINORIDENS Melvill 1901 = valid, 65
minuta Gmelin 1791 = gangranosa
misella Perry 1811 = ziczac
modesta Sowerby 1870 = owenii
moelleri Iredale 1931 = angustata
momokiti Steadman & Cotton 1943 = eglantina
MONETA Linnaeus 1758 = valid, 149
monetoides Iredale 1939 = moneta
moniliaris Lamarck 1810 = clandestina
moniontha Melvill 1888 = stolida
monochroma Mollerat 1890 = lurida
monstrans Iredale 1935 = limacina
monstrosa Gray 1828 = lurida
monstrosa Sowerby 1870 = lurida
monstrosa Smith 1878 = ?esontropia
montosa Roberts 1870 = mappa
montrouzieri Dautzenberg 1903 = mappa
morbillosa Roeding 1798 = chinensis
moretonensis Schilder 1965 = langfordi
mozambicana Schilder & Schilder 1938 = nebrites
multidentata Coen 1949 = caurica
murina Nardo 1847 = lurida
MUS Linnaeus 1758 = valid, 116
musumea Kuroda & Habe 1961 = katsuae

nana Crosse 1896 = achatidea
nandronga Steadman & Cotton 1943 = stolida
nariaeformis Schilder 1930 = albuginosa
nasese Steadman & Cotton 1943 = labrolineata
nashi Iredale 1931 = labrolineata
NEBRITES Melvill 1888 = valid, 160
nebulosa Gmelin 1791 = stercoraria
nebulosa Kiener 1843 = zonaria
nebulosa Monterosato 1897 = lurida
nebulosa Solander 1786 = mauritiana
neglecta Sowerby 1837 = hirundo
nephelodes Lancaster 1928 = tigris

nigella Sulliotti 1924 = scurra
niger Roberts 1885 = eglantina
nigrescens Sulliotti 1924 = stercoraria
nigrescens Gray 1824 = tigris
nigricans Crosse 1869 = eglantina
nigricans Crosse 1875 = mappa
nigricans Dautzenberg 1923 = tigris
nigricans Pallary 1926 = caurica
nigrocincta Schilder 1924 = caurica
nigroguttata Coen 1949 = lynx
NIGROPUNCTATA Gray 1828 = valid, 190
nimbosa Dillwyn 1827 = quadrimaculata
nimiserrans Iredale 1835 = errones
nitida Coen 1949 = fimbriata, gracilis, staphylaea
nitens Coen 1949 = caurica, semiplota
nitidiuscula Coen 1949 = spurca
nivea Gray 1824 = turdus
nivea Wood 1828 = lutea
nivea Preston 1909 = miliaris
NIVOSA Broderip 1827 = valid, 172
nivosa Coen 1949 = pyrum
normalis Monterosato 1897 = lurida, pyrum, spurca
northi Steadman & Cotton 1943 = cribraria
notata Gill 1858 = gracilis
noumeensis Marie 1869 = annulus
novaebritanniae Schilder & Schilder 1937 = pulchella
novaecaledoniae Schilder & Schilder 1952 = childreni
nubilis Sulliotti 1924 = annulus
NUCLEUS Linnaeus 1758 = valid, 205
nugata Iredale, 1935 = saulae
nukulau Steadman & Cotton 1943 = staphylaea
numisma Roeding 1798 = moneta
nymphae Jay 1850 = onyx

obesa Fulton 1930 = gangranosa
oblonga Gmelin 1791 = errones
oblongata Melvill 1888 = caurica
obscura Crosse 1869 = caurica
obscurata Schilder & Schilder 1940 = caurica
obstructa Coen 1949 = lurida
obtusa Perry 1811 = pantherina
obvelata Lamarck 1810 = annulus
occidentalis Iredale 1935 = bicolor
oceanica Schilder 1930 = lurida
OCELLATA Linnaeus 1758 = valid, 217
ochroleuca Gmelin 1791 = unrecognizable
oculata Roeding 1798 = histrio
ogasawarensis Schilder 1944 = cernica
ogasawarensis Schilder 1944 = hirundo
olivacea Gmelin 1791 = stercoraria
olivacea Lamarck 1810 = ovum
onca Roeding 1798 = tigris
onycina Coen 1949 = lurida
ONYX Linnaeus 1758 = valid, 106
oranica Crosse 1896 = achatidea
orcina Iredale 1931 = vitellus
orientalis Schilder & Schilder 1940 = cribraria
ostergaardi Dall 1921 = thomasi
otahitensis Schubert & Wagner 1829 = ventriculus
ovata Gmelin 1791 = ?lynx
ovata Gmelin 1791 = mauritiana
ovata Gray 1824 = ovum
ovata Perry 1811 = turdus
OVUM Gmelin 1791 = valid, 138
ovum Roeding 1798 = mus
OWENII Sowerby 1837 = valid, 88

pacifica Ostergaard 1920 = thomasi
pacifica Steadman & Cotton 1943 = lynx
palatha Melvill 1888 = ocellata
palauensis Schilder & Schilder 1938 = ovum
pallens Taylor 1916 = erosa
pallida Pallary 1900 = achatidea, pyrum
PALLIDA Gray 1824 = valid, 187
pallida Dautzenberg 1903 = caurica, eglantina
pallidior Dautzenberg 1903 = errones
PALLIDULA Gaskoin 1849 = valid, 247
panerythra Melvill 1888 = mappa
pantherina Pallary 1938 = spurca
PANTHERINA Solander 1786 = valid, 169
pardalina Dunker 1852 = turdus
pardalis Shaw 1795 = tigris
pardalis Roeding 1798 = tigris
pardus Roeding 1798 = pantherina
parvula Philippi 1849 = ?hirundo
paschalis Roeding 1798 = mauritiana
passerina Melvill 1888 = clandestina
pauciguttata Schilder & Schilder 1938 = felina
peasei Sowerby 1870 = gaskoini
pelidna Melvill & Standen 1904 = ocellata
pellucens Melvill 1888 = teres
pellucida Taylor 1916 = esontropia
pellucidula Sulliotti 1924 = cinerea
pentella Iredale 1939 = teres
percomis Iredale 1931 = cernica
perconfusa Iredale 1935 = eglantina
pericalles Melvill & Standen 1904 = pulchella
perlacea Sulliotti 1924 = erosa
perlae Lopez & Chiang 1975 = decipiens
peropima Iredale 1939 = hirundo
perrieri Rochebrune 1884 = annulus
persica Schilder & Schilder 1938 = onyx
persticta Iredale 1939 = punctata
petechialis Roeding 1798 = sanguinolenta
petitiana Crosse 1872 = pyrum
phagedaina Melvill 1888 = erosa
phyllidis Shaw 1915 = turdus
physoides Coen 1949 = pyrum
physoides Monterosato 1897 = achatidea
picta Gray 1824 = zonaria
PIPERITA Gray 1825 = valid, 135
piperitoides Coen 1949 = pyrum
piriformis Locard 1886 = pyrum
piscatorum Schilder 1965 = ?subviridis
pleuronectes Rochebrune 1884 = moneta
plumbea Gmelin 1791 = zebra
polita Roberts 1868 = semiplota
polynesiae Schilder & Schilder 1939 = vitellus
PORARIA Linnaeus 1758 = valid, 221
PORTERI Cate, 1966 = valid, 228
prasina Shaw 1909 = arabica
pressa Dillwyn 1827 = annulus
prestoni Shaw 1909 = interrupta
pretiosa Coen 1949 = carneola
pretiosa Melvill 1905 = ocellata
princeps Gray 1824 = valentia
proba Iredale 1939 = errones
problematica Iredale 1935 = pyriformis
prodiga Iredale 1939 = cernica
propinqua Garrett 1879 = carneola
protracta Dautzenberg 1906 = erosa
prunus Roeding 1798 = onyx
pseudarabicula Coen 1949 = caurica
pseudocellata Schilder & Schilder 1938 = marginalis
pseudovinosa Schilder 1927 = camelopardalis
pubescens Monterosato 1897 = lurida
pudica Rous 1905 = zebra
pulchella Gray 1824 = pulchra

pulchella **Gray** 1828 = owenii
pulchella Coen 1949 = erosa
PULCHELLA Swainson 1823 = valid, 162
PULCHRA Gray 1824 = valid, 162
pulchroides Alvarado & Alvarez 1964 = lurida
PULICARIA Reeve 1846 = valid, 207
pulla Gmelin 1791 = onyx
pumilio Brusina 1865 = lurida
PUNCTATA Linnaeus 1771 = valid, 73
punctata Pallary 1900 = achatidea
punctatissima Coen 1949 = caputdraconis
punctulata Gray 1824 = robertsi
punctulata Gmelin 1791 = ?lynx
purissima Vredenburg 1919 = erosa
purissima Coen 1949 = cicercula
purpurascens Gmelin 1791 = unrecognizable
purpurascens Swainson 1823 = sanguinolenta
pusilla Gmelin 1791 = ?errones
PYRIFORMIS Gray 1824 = valid, 174
pyriformis Sowerby 1870 = turdus
PYRUM Gmelin 1791 = valid, 115

QUADRIMACULATA Gray 1824 = valid, 147
quadrimaculata Dautzenberg & Bouge 1933 = bistrinotata
queenslandica Schilder 1966 = hirasei
quinquefasciatus Roeding 1798 = caurica

RABAULENSIS Schilder 1964 = valid, 245
raripunctata Sulliotti 1911 = turdus
RASHLEIGHANA Melvill 1888 = valid, 211
rattus Lamarck 1810 = stercoraria
raysummersi Schilder 1960 = hammondae
redimita Melvill 1888 = lamarckii
reductesignata Schilder 1924 = kieneri
reentsii Dunker 1852 = gangranosa
REEVEI Sowerby 1832 = valid, 126
regina Gmelin 1791 = mauritiana
reticulata Gmelin 1791 = histrio
reticulata Chenu 1845 = maculifera
reticulatum Gmelin 1791 = caputserpentis
reticulifera Schilder 1924 = bicolor
retifera Menke 1829 = scurra
rewa Steadman & Cotton 1943 = mappa
rhinoceros Souverbie 1865 = pallidula
rhomboides Schilder & Schilder 1933 = moneta
ROBERTSI Hidalgo 1906 = valid, 154
rosea Gray 1824 = mappa
rosea Gray 1825 = moneta
rosea Taylor 1913 = caurica
ROSSELLI Cotton 1948 = valid, 110
rossiteri Dautzenberg 1903 = tigris, walkeri
rostrata Dautzenberg 1903 = annulus, cribraria, punctata
rouxi Ancey 1882 = hirundo
rubiginosa Gmelin 1791 = stolida
rubiola Kenyon 1902 = carneola
rufa Lamarck 1810 = pyrum
rufa Coen 1949 = eglantina
rufescens Costa 1829 = lurida
rufescens Gmelin 1791 = cinerea
rufodentata Raybaudi 1978 = stolida
rufofulva Kobelt 1906 = lurida
rufula Mollerat 1890 = achatidea
rumphii Schilder & Schilder 1938 = isabella
russonitens Melvill 1888 = tigris
ruvaya Steadman & Cotton 1943 = limacina

SAKURAI Habe 1970 = valid, 256
salita Roberts 1870 = vitellus
samurai Schilder & Schilder 1940 = childreni
sanctaehelenae Schilder 1930 = spurca

SANGUINOLENTA Gmelin 1791 = valid, 85
sarcodes Melvill 1888 = vitellus
saturata Dautzenberg 1903 = talpa
SAULAE Gaskoin 1843 = valid, 82
scarabaeus Bory 1827 = poraria
schilderiana Cate 1961 = tigris
SCHILDERORUM Iredale 1939 = valid, 64
schneideri Schilder & Schilder 1938 = kieneri
scotti Broderip 1831 = friendii
SCURRA Gmelin 1791 = valid, 216
scutellum Schilder & Schilder 1937 = annulus
SEMIPLOTA Mighels 1845 = valid, 212
senegalensis Schilder 1928 = pyrum
SERRULIFERA Schilder & Schilder 1938 = valid, 68
siasiensis Cate 1960 = saulae
siciliana van Salis 1793 = pyrum
signata Iredale 1939 = ziczac
similis Gmelin 1791 = erosa
similis Gray 1831 = fuscorubra
simulans Schilder & Schilder 1940 = pallidula
sista Iredale 1939 = cylindrica
smithi Sowerby 1881 = pyriformis
sophia Brazier 1897 = caputserpentis
sophiae Brazier 1876 = ovum
sordida Lamarck 1810 = cinerea
sorrentensis Schilder 1963 = venusta
sosokoana Ladd 1934 = annulus
sowerbyana Schilder 1932 = cylindrica
sowerbyi Anton 1839 = carneola
sowerbyi Kiener 1845 = annettae
SPADICEA Swainson 1823 = valid, 127
spadix Mighels 1845 = semiplota
sphaeridium Schilder & Schilder 1938 = globulus
splendens Gyngell 1924 = arabica
splendens Taylor 1916 = chinensis
SPURCA Linnaeus 1758 = valid, 182
squalina Gmelin 1791 = lynx
standeni Melvill 1905 = scurra
STAPHYLAEA Linnaeus 1758 = valid, 213
steineri Cate 1969 = coxeni
stellata Gmelin 1791 = helvola
stellata Perry 1811 = ?nebrites
STERCORARIA Linnaeus 1758 = valid, 173
stercusmuscarum Lamarck 1810 = punctata
stohleri Cate & Schilder 1968 = contaminata
STOLIDA Linnaeus 1758 = valid, 157
straminea Melvill 1888 = erosa
striata Gmelin 1791 = helvola
striatella Link 1807 = asellus
sturangi Schilder & Schilder 1938 = nucleus
subalata Schilder & Schilder 1933 = moneta
subalba Smith 1912 = nebrites
subcarnea Beddome 1896 = angustata
subcoerulea Schilder & Schilder 1931 = gracilis
subcylindrica Sowerby 1870 = cylindrica
subfasciata Link 1807 = teres
subflava Gmelin 1791 = lynx
subfossilis Schilder 1930 = cinerea
subfuscula Dillwyn 1827 = vitellus
sublaevis Schilder & Schilder 1938 = bistrinotata
subrostrata Dautzenberg 1903 = vitellus
subsignata Melvill 1888 = mappa
SUBTERES Weinkauff 1881 = valid, 109
subulata Dautzenberg 1903 = hirundo
SUBVIRIDIS Reeve 1835 = valid, 163
succincta Linnaeus 1758 = onyx
SULCIDENTATA Gray 1824 = valid, 105
SUMMERSI Schilder 1958 = valid, 95
superstes Schilder 1930 = martini
supertecta Sulliotti 1924 = camelopardalis
surabajensis Schilder 1937 = walkeri
SURINAMENSIS Perry 1811 = valid, 175
surinensis Raybaudi 1978 = guttata
suta Coen 1949 = mus
suvaensis Steadman & Cotton 1943 = fimbriata
sydneyensis Schilder 1938 = chinensis
syringa Melvill 1888 = pantherina
tabescens Dillwyn 1817 = teres
TALPA Linnaeus 1758 = valid, 233
tectoriata Sulliotti 1924 = annulus
TERAMACHII Kuroda 1938 = valid, 232
TERES Gmelin, 1791 = valid, 219
TESSELLATA Swainson 1822 = valid, 195
TESTUDINARIA Linnaeus 1758 = valid, 164
testudinosa Perry 1811 = testudinaria
tetsuakii Kira 1959 = globulus
TEULEREI Cazenavette 1846 = valid, 239
thakau Steadman & Cotton 1943 = stolida
thatcheri Cox 1869 = venusta
theeva Steadman & Cotton 1943 = dillwyni
thema Iredale 1939 = caurica
theoreta Iredale 1939 = poraria
thepalea Iredale 1939 = carneola
theriaca Melvill 1888 = pantherina
thersites Gaskoin 1849 = friendii
thielei Schilder & Schilder 1938 = quadrimaculata
THOMASI Crosse 1865 = valid, 96
tigrina Gmelin 1791 = tigris
tigina Lamarck 1822 = pantherina
TIGRIS Linnaeus 1758 = valid, 176
timorensis Gray 1825 = ?cicercula
timorensis Kenyon 1902 = helvola
titan Schilder & Schilder 1962 = carneola
tomlini Schilder 1930 = cernica
topee Steadman & Cotton 1943 = ventriculus
tortirostris Sowerby 1906 = chinensis
translucens Gmelin 1791 = cinerea
translucida Melvill 1888 = esontropia
transpiciens Taylor 1916 = teres
trenberthae Trenberth 1961 = piperita
tricornis Jousseaume 1874 = margarita
trifasciata Monterosato 1897 = pyrum
trifasciata Gmelin 1791 = mauritiana
trigonella Blainville 1826 = tessellata
tristis Sulliotti 1924 = tigris
trizonata Sowerby 1870 = punctata
tuberculata Gray 1828 = mus
tuberculifera Sulliotti 1924 = tigris
tuberculigera Sulliotti 1924 = carneola
tuberculosa Quoy & Gaimard 1834 = moneta
turanga Steadman & Cotton 1943 = aurantium
turbinata Gmelin 1791 = mauritiana
turdiculus Monterosato 1897 = lurida
TURDUS Lamarck 1810 = valid, 114
typica Smith 1903 = arabica
typica Monterosato 1897 = pyrum
umbilicata Dillwyn 1823 = onyx
umbilicata Sowerby 1825 = armeniaca
undata Lamarck 1810 = ziczac
undata Pallary 1900 = pyrum
undosa Roeding 1798 = histrio
undulata Gmelin 1791 = mauritiana
undulata Wood 1828 = ziczac
unifasciatus Mighels 1845 = fimbriata
URSELLUS Gmelin 1791 = valid, 80

VALENTIA Perry 1811 = valid, 125
vellei Jaume & Borro 1946 = zebra
vanelli Linnaeus 1758 = lynx

variolaria Lamarck 1810 = chinensis
variolosa Gmelin 1791 = pyrum
variolosa Roeding 1798 = nucleus
vasta Schilder & Schilder 1938 = owenii
vaticina Iredale 1931 = subviridis
vatu Steadman & Cotton 1943 = felina
vava Steadman & Cotton 1943 = teres
vayssierei Schilder & Schilder 1938 = pulchella
velesia Iredale 1939 = felina
venerea Gmelin 1791 = mauritiana
ventricosa Gray 1824 = argus
ventricosa Mollerat 1890 = lurida
VENTRICULUS Lamarck 1810 = valid, 122
VENUSTA Sowerby 1846 = valid, 120
vercoi Schilder 1930 = friendii
verconis Cotton & Godfrey 1932 = angustata
verdensium Melvill 1888 = spurca
vespa Donovan 1820 = asellus
vespacea Melvill 1905 = asellus
vestimenti Rochebrune 1884 = moneta
vibex Kenyon 1902 = poraria
vinosa Gmelin 1791 = pantherina
violacea Rous 1905 = chinensis
virescens Hidalgo 1907 = lurida
virginalis Schilder & Schilder 1938 = diluculum
viridacea Sulliotti 1924 = pulchra
viridicolor Cate 1962 = cernica
viridis Kenyon 1902 = mappa
viridis Sulliotti 1924 = stercoraria
viridosa Sulliotti 1924 = scurra
viridula Sulliotti 1924 = spurca
VITELLUS Linnaeus 1758 = valid, 254
vitellus Gray 1825 = **annulus**

vitiensis Steadman & Cotton 1943 = kieneri
vittata Deshayes 1831 = **ziczac**
vivia Steadman & Cotton 1943 = **pallidula**
vivili Steadman & Cotton 1943 = **errones**
volai Steadman & Cotton 1943 = **tigris**
vono Steadman & Cotton 1943 = **scurra**
VREDENBURGI Schilder 1927 = valid, 78
vulavula Steadman & Cotton 1943 = globulus

waikikiensis Schilder 1933 = fimbriata
WALKERI Sowerby 1832 = valid, 150
wangga Steadman & Cotton 1943 = cylindrica
westralis Iredale 1935 = histrio
whitleyi Iredale 1939 = clandestina
whitworthi Cate 1964 = chinensis
wilhelmina Kenyon 1897 = poraria
wilkinsi Griffiths 1959 = bicolor
williamsi Melvill 1888 = lynx
winckworthi Schilder & Schilder 1938 = turdus

xanthochrysa Melvill 1888 = sulcidentata
XANTHODON Sowerby 1832 = valid, 167

yaloka Steadman & Cotton 1943 = humphreysi

zadela Iredale 1939 = cribraria
zanzibarica Sulliotti 1911 = turdus
ZEBRA Linnaeus 1758 = valid, 189
ZICZAC Linnaeus 1758 = valid, 69
ZONARIA Gmelin 1791 = valid, 179
zonata Lamarck 1810 = zonaria
zonata Mollerat 1890 = lurida
zymecrasta Melvill 1888 = tigris